"I am impressed each time by the anci
presented by Aidan Rankin. The climat
nerability of human life when it become:
origin. Therefore, the necessity of brid{
ern world and the ancient cultures is mo.. ...
— *Fons Elders, Emeritus Professor of the Theory of Worldview...
at the University for Humanist Studies, Utrecht, the Netherlands*

"This book fills an important gap in our understanding of the connection between Jainism (and other Dharmic traditions influenced by Jainism) and environmental politics. It emphasises the important, perhaps essential, role that non-western worldviews can play in addressing issues of sustainability. The book is very timely in that it coincides with the rise of India as an economic power facing severe environmental problems as well as a greater global interest in the role of spiritual or religious traditions in addressing environmental questions. It is an important read for environmental activists seeking to incorporate non-western ideas into their political thinking."
—*Daniel T. Ostas, Professor & James G. Harlow, Jr. Chair
in Business Ethics, Michael F. Price College of Business,
University of Oklahoma, USA*

"Impressive and fascinating insights into Jainism. The case study of the jeweller Vardhaman Gems and its business and environmental philosophies illustrates succinctly how Jain principles are applied successfully in a practical way."
—*Paul Barker, President, Leeds Theosophical Society, UK*

"Through this book, Aidan Rankin open us up to the breadth and depth of Jainism with its outlook and practices that are deeply relevant for our time of ecological crisis. This book brings much needed hope in these challenging times."
—*Ian Mowll, Interfaith Minister (UK) and
Coordinator of GreenSpirit (www.greenspirit.org.uk)*

Jainism and Environmental Politics

This book explores the ways in which the ecologically centred Indian philosophy of Jainism could introduce a new and non-western methodology to environmental politics, with the potential to help the green movement find new audiences and a new voice.

Aidan Rankin begins with a description of the ideas and principles that distinguish Jainism from other Indian (and western) philosophies. He goes on to compare and contrast these principles with those of current environmental politics and to demonstrate the specific ways in which Jain ideas can assist in driving the movement forward. These include the reduction of material consumption, the ethical conduct of business within sustainable limits, and the avoidance of exploitative relationships with fellow humans, animals and ecosystems. Overall, the book argues that Jain pluralism could be a powerful tool for engaging non-western societies with environmental politics, allowing for an inclusive approach to a global ecological problem.

This book will be of great interest to students and scholars of environmental politics, environmental philosophy, comparative religions and Jainism.

Aidan Rankin is an author, independent scholar and property consultant based in London, UK. He has a PhD in Political Science from the London School of Economics and an MA in Modern History from the University of Oxford.

Routledge Focus on Environment and Sustainability

Climate Adaptation Finance and Investment in California
Jesse M. Keenan

Negotiating the Environment: Civil Society, Globalisation and the UN
Lauren E. Eastwood

Carbon Inequality
The Role of the Richest in Climate Change
Dario Kenner

The UNESCO Manual for Developing Intercultural Competencies
Story Circles
Darla K. Deardorff

Design for Sustainability
A Multi-level Framework from Products to Socio-technical Systems
Fabrizio Ceschin and İdil Gaziulusoy

Sustainability, Conservation and Creativity
Ethnographic Learning from Small-scale Practices
Pamela J. Stewart and Andrew J. Strathern

Jainism and Environmental Politics
Aidan Rankin

For more information about this series, please visit: https://www.routledge.com/Routledge-Focus-on-Environment-and-Sustainability/book-series/RFES

Jainism and Environmental Politics

Aidan Rankin

LONDON AND NEW YORK

from Routledge

First published 2020
by Routledge
2 Park Square, Milton Park, Abingdon, Oxon OX14 4RN

and by Routledge
52 Vanderbilt Avenue, New York, NY 10017

Routledge is an imprint of the Taylor & Francis Group, an informa business

First issued in paperback 2021

© 2020 Aidan Rankin

The right of Aidan Rankin to be identified as author of this work has been asserted by him in accordance with sections 77 and 78 of the Copyright, Designs and Patents Act 1988.

All rights reserved. No part of this book may be reprinted or reproduced or utilised in any form or by any electronic, mechanical, or other means, now known or hereafter invented, including photocopying and recording, or in any information storage or retrieval system, without permission in writing from the publishers.

Trademark notice: Product or corporate names may be trademarks or registered trademarks, and are used only for identification and explanation without intent to infringe.

British Library Cataloguing-in-Publication Data
A catalogue record for this book is available from the British Library

Library of Congress Cataloging-in-Publication Data
Names: Rankin, Aidan, author.
Title: Jainism and environmental politics / Aidan Rankin.
Description: 1. | New York: Routledge, 2020. |
Series: Routledge focus on environment and sustainability |
Includes bibliographical references and index.
Identifiers: LCCN 2019038532 (print) | LCCN 2019038533 (ebook) |
ISBN 9780367189761 (hardback) | ISBN 9780367189778 (ebook)
Subjects: LCSH: Ecology—Religious aspects—Jainism. |
Environmentalism—Religious aspects—Jainism. |
Environmentalism—Political aspects.
Classification: LCC BL1375.E36 R37 2020 (print) |
LCC BL1375.E36 (ebook) | DDC 294.4/177—dc23
LC record available at https://lccn.loc.gov/2019038532
LC ebook record available at https://lccn.loc.gov/2019038533

ISBN: 978-0-367-18976-1 (hbk)
ISBN: 978-1-03-208492-3 (pbk)
ISBN: 978-0-367-18977-8 (ebk)

Typeset in Times New Roman
by codeMantra

For Brian

Contents

Acknowledgements		*x*
Foreword by Lynne Sedgmore		*xii*
Glossary		*xiv*
1	Introduction	1
2	The Jain voice and political ecology	11
3	The Jain theory of pluralism: transcending the politics of protest?	37
4	Jainism and environmental politics: a radical synthesis?	53
	Index	*79*

Acknowledgements

I am not a Jain, and so I look upon this ancient Indian philosophical and cultural tradition with an outsider's curiosity and respect. My mother, Anne Vannan Rankin, first taught me the principle of compassion for all forms of life, which is, as I later discovered, the first principle of Jain doctrine. My father, Professor David Rankin, has also been a constant source of wise advice. I have also been fortunate in that many Jains have encouraged my interest and facilitated my research. Professor Kanti Mardia of Leeds and Oxford universities has taught me to look for hidden connections between ancient spiritual insights and modern scientific discoveries. He, his wife Pavan and their extended family are exemplars of Jain hospitality. They have made me welcome at their home in Leeds and at the Yorkshire Jain Community, to which they have devoted many years of work.

Atul Shah, my former colleague at the London School of Economics, stimulated my early interest in Jain thought when he edited *Jain Spirit* magazine, for which I was invited to serve as Guest Editor, an exceptional honour for a non-Jain. Atul opened my mind to many of Jainism's hidden depths and subtleties, always emphasising their relevance to current political and social questions, not least the mounting ecological crisis. Mahersh and Nishma Shah, of Shambhu's catering, are models of humane business practice warmly supported me in my efforts to understand the ethical system that governs their commercial practice and charitable endeavours. Mark Sattin has explored with me in considerable depth the experience of the California Greens and the relevance of the concept of the 'radical middle' to a many-sided understanding of politics that accords with Jain values. I greatly value his friendship and generous feedback. Less directly, I have profoundly benefitted from Jyoti Kothari's extensive knowledge and fascinating exploration of the extensive Jain contribution to India's jewellery trade.

Acknowledgements xi

At Routledge, I would like to thank my Commissioning Editor, Annabelle Harris, for her encouragement and wise advice throughout the writing of this and my previous book, *Jainism and Environmental Philosophy*. I have also been extremely lucky to have worked for a second time with Matthew Shobbrook as my Editorial Assistant. I am also once again very grateful to Mark Wells for his excellent and indispensable work as indexer. Lynne Sedgmore, CBE, has been a source of knowledge, support and friendship. I am especially grateful to her for contributing the Foreword to this book.

Finally, and most important of all, Brian Scoltock has put up with me patiently throughout the writing of this book: his belief in the project has been crucial to me at all times.

Foreword

In these times of increasing awareness of the perilous plight of our planet, Aidan's book is a refreshing and powerful analysis of the invaluable contribution of Jain philosophy and spiritual practice to the various issues facing the world today.

I have been fascinated by Jainism since the early 1990s, and Aidan's fascinating books (six in all) have enabled me, a white woman brought up in the Christian tradition, living in the West, to delve deep into all aspects and facets of this graceful, profound and inspiring faith tradition.

This book has the civilian Jain as its focal point – the man or woman whose daily spiritual practice incorporates acting with respect and care for their natural environment. Such individuals continually consider the impact of their current actions on future generations. If everyone tried to live that way, could significant change happen overnight to save Gaia?

Aidan offers us a treasure trove from an ancient and sustainable tradition whose wisdom is highly pertinent to our troubled modern world. We live in societies that harm and dominate our beautiful planet for selfish ends; we live at the edge of potentially disastrous climate change and biodiversity extinction.

It is time to learn from the Jain philosophy and practices of non-injury, peaceful living, philanthropy, pluralism, interconnectedness, balance, moderation, cooperation, responsibility and social justice – not only for humans but for all species.

My favourite Jain wisdom – many sidedness or Anekant, the ability to see and accept alternative viewpoints – is explored. I remain convinced that this is a practice desperately needed to inform and transform our increasingly polarised politics and rising violent intolerance.

Aidan Rankin does not romanticise the Jains, and yet he is clearly an admirer of them as 'green pioneers', with a special contribution from the Global South. His robust articulation of the differences between the thinking of Jains and modern green and eco-movements is a fascinating delight awaiting each reader. Enjoy: I certainly did.

I believe this book will intrigue, inform and inspire a wide audience – the newcomer to Jainism, anyone interested in wisdom steeped in ecological and green awareness, students of ecology and faith traditions, eco-activists and political influencers.

Who knows, it might even contribute to saving our precious planet.

Lynne Sedgmore, CBE
Former Chief Executive,
Centre for Excellence in Leadership.
Interfaith Minister, Poet and Leadership Coach.
Named in Debretts 500 List of top UK influencers 2015 and UK 100 Women of Spirit 2016.

Glossary

The Jain tradition has a rich vocabulary, drawn both from the classical *Sanskrit* language and the vernacular *Prakrit*, known to linguistic scholars as *Ardhamagadhi*. The list below covers the concepts and expressions used in this study. In the interests of clarity, accents and diacritical marks have not been applied below or in the main text because there is no common agreement on when or how they should be applied. Diacritical marks and accents are, however, reproduced when quoting from other texts and when those texts are cited in the chapter bibliographies. In the main text, some Sanskrit or Prakrit words are written with capital letters, with others written in the lower case. This is based both on convention and the original sources in which these terms appear.

All Sanskrit and Prakrit words and phrases used in this book are written in italics apart from the following: Ahimsa; Anekantavada (or Anekant); Dharma; jina, and jiva. This is because these words express key concepts associated with this study and so are used more frequently than other Jain terminology,

Ahimsa Non-Violence, non-injury, avoidance of harm.
Ajiva Inert matter, insentient object without jiva (life monad or 'soul').
Akriyavada Doctrine of non-action.
Alparambhi Requiring minimum violence or harm: used mainly to describe occupations and trades.
Anekantavada Principle of 'many-sidedness' (of reality), pluralism or 'multiple viewpoints' (also **Anekant**, **Anekanta**, **Anekantvada**).
Anekantavadin(s) Adherent or proponent of **Anekantavada**.
Anga(s) Central Jain text(s), (literally 'limbs' of the Jain canon).
Anityavada Naya of non-eternalism (the idea that no substance (**Dravya**) or material form exists in a permanent form). Opposite

of **Nityavada**. (The Jain position is that substance exists in a permanent form but is also subject to change.)
Anu   Infinitesimal or sub-atomic particle.
Anuvrata(s)   Five Lesser Vows undertaken by laymen and laywomen.
Aparigraha   Principle of non-possessiveness.
Asrava   Beginning of 'karmic bondage' (**Bandha**), influx of karmic particles.
Asteya   (sometimes called **Achaurya**): 'Non-stealing', avoidance of theft or taking what is not given.
Astikya   Implicit understanding of the nature of reality.
Avasarpini   Regressive half-cycle of time and the universe (complemented by **Utsarpini**).
Avidya   Ignorance, lack of knowledge or perception.
Bandha   Karmic bondage.
Brahmacharya   Celibacy (for ascetics), fidelity and avoidance of promiscuity or exploitative relationships (for laymen and laywomen).
Charvaka (ucchedavada)   **Naya** of 'annihilationism' or belief that only finite material things exist.
Darshana   Perception, (clear) vision.
Dharma   Universal law or cosmic order, encompassing philosophy and religion.
Digambar (or Digambara)   One of the two main schools of Jainism: literally means 'sky-clad' because the most senior male ascetics are naked.
Dravya   Substance.
Ekant (Ekantika)   One-sidedness, doctrinaire viewpoint.
Ekantavadin   Adherent or proponent of **Ekant**.
Eva   'In fact' conveys subjective 'fact' in **Syadvada**, i.e. perceived fact according to speaker.
Gyana (or Jnana)   Knowledge.
Himsa   Violence, harm, destructive power.
Irya-Samiti   Principle of Careful Action.
Jai Jinendra   'Honour to [the] **Jina**', 'Hail to the Conqueror(s)', popular Jain greeting.
Jain-ness (Jainness)   Cultural sensibility shared by Jains.
Jina(s)   Spiritual victor(s), omniscient spiritual teacher(s).
Jiva   Life monad or 'soul'.
Jiva Daya   Sympathy or identification with all sentient beings.
Karma   (in Jain thought) Subtle matter composed of karmic particles, attracted to the Jiva by **Yoga** (activity) and preventing full self-knowledge, omniscience and transcendence of *samsara* (the cycle of birth, death and rebirth).

Glossary

Karma pudgala Karmic matter.
Karmon(s) Karmic particle(s): **Karmons** are also sometimes referred to as '**Karmas**'.
Kasaya Passion(s), usually negative, such as anger, fanaticism or material greed
Lokakasa Inhabited universe, occupied space.
Mahavrata(s) Five Greater Vows undertaken by ascetic men and women.
Maya Deceit, illusion.
Mithyatva (Mithyadarshana) False consciousness, distorted worldview.
Moksha Spiritual liberation, enlightenment, acquisition of omniscience.
Muni Ascetic man or woman.
Naya Viewpoint, partial truth.
Nirjara Breakage, shedding, falling away of karma/karmic particles.
Nirvana Full enlightenment, moment of attaining enlightenment or becoming a **siddha** or liberated soul.
Nityavada **Naya** of eternalism (the idea that some aspects of existence remain eternally unchanging).
Niyativada **Naya** of fatalism, including most forms of theism.
Panjrapoor Animal hospital or sanctuary.
Papa Negative, destructive or 'heavy' karma.
Parasparopagraho Jivanam Concept of interconnectedness. Philosophical translation: All life is bound together by mutual support and interdependence; Religious translation: Souls (jiva) render service to one another.
Parigraha Possessiveness.
Pudgala Matter.
Punya Positive, benevolent, creative or 'light' karma
Ratnatraya or Triratna (Tri-ratna) Three Jewels of Jainism: **Samyak Darshana** (Right Faith); **Samyak Gyana (or Jnana)** (Right Knowledge) and **Samyak Charitra** (Right Action/Conduct).
Samiti Rules of conduct for ascetics.
Samsara Cycle of birth, death and rebirth, process of cyclic change.
Samvara Stoppage of karmic influx (through awakening consciousness).
Samyak Charitra Right Conduct.
Samyak Darshana (Samyaktva) Right Faith.
Samyak Gyana (or Jnana) Right Knowledge.
Sarvodaya Principle of 'compassion for all' universal uplift.
Satya Truth, truthfulness, honesty.
Shraddha Educated faith, intuition reinforced by knowledge and reason

Shuksha Education.
Siddha Liberated soul(s).
Stotra(s) Series of poetic hymns to the **Jinas**.
Sutra Collection of spiritual aphorisms or teachings, usually in verse form ('thread' or 'string' in Sanskrit).
Svetambar (or Svetambara) One of the two main schools of Jainism literally means 'white-clad' because male and female ascetics wear white robes.
Swadeshi Principle and practice of economic self-sufficiency or self-reliance (popularised by Gandhi and not exclusive to Jains): literally means 'of one's own country'.
Syadvada System of logic based on qualified definition.
Syat Expression of possibility in **Syadvada**: literally 'might be'.
Tattva(s) 'Nine Reals', aspects of reality or 'things'.
Tirtha Ford to be crossed, to which **Samsara** is compared.
Tirthankara Ford-maker or guide to enlightenment, omniscient spiritual teacher. Most elevated form of **Jina**.
Utsarpini Progressive half-cycle of time and the universe (complemented by **Avasarpini**).
Vrata(s) Vow, vows.
Yoga Activities of body, mind and speech.

1 Introduction

At first glance, the connections between the Indian faith tradition known to westerners as 'Jainism' and the politics of the environment seem so apparent that they barely merit discussion. Jains follow a strictly vegetarian or in some cases vegan diet. Their philosophy of life is rooted in non-violence, the conscious avoidance of injury or harm to all forms of life and the connections between all living systems, humans included. Moreover, Jains recognise that even the most minuscule life forms (from a human perspective) are complex, intricate beings. They can also play vital roles in the maintenance of a habitable planet: their survival is intertwined with our survival. Jains therefore construct their lives around the principle of inflicting minimal harm on all living creatures. A small minority of ascetics, male and female, practise extreme austerities to avoid injury to life. 'Civilian' Jains, by contrast, live as peacefully as possible and take conscious steps to foreswear unnecessary or conspicuous consumption. They also are enjoined at all times to exercise care to ensure that other species, including plant species, are not unnecessarily harmed or even disturbed. The Jain commitment to social justice includes fellow humans, with exploitative relationships of all kinds viewed as harmful to exploiters and exploited alike. More radically, it extends the concept of 'social' to include non-human species. Society consists of far more than merely human communities and the idea of a humanity detached from nature by superior intelligence is viewed as a dangerous delusion of grandeur. Human intelligence is measured by the exercise of restraint rather than dominance, human knowledge by an awareness of our limitations (individual and collective) and our dependence on the natural world.

These insights have been acquired and refined over millennia, for Jainism is one of the world's oldest faiths and earliest philosophical traditions. Its modern adherents regard themselves as the inheritors of the most ancient forms of Indic thought. The intuitions of a nonliterate society pointed towards a world, and beyond that a universe, teeming with life. The life principle itself is embedded in every type of being, with all forms of life serving their unique purpose but also mutually dependent. Insights such as these are based on a sense of enchantment with the natural world at one level, balanced by a sense of vulnerability to natural or cosmic forces. Such intuitions have been overlaid by many layers of scientific and philosophical speculation, including openness to viewpoints from other cultures. There is a critical openness, for example, towards western thought and the many schools of Hinduism, the majority faith tradition that surrounds Jain communities in India. The Jain tradition is enlightened, rational and pluralist. It can also be seen as a cultural sensibility, based on instinctive understanding and a spiritual or devotional practice. Out of these two 'branches' of Jain thought, a rich culture and an intricate philosophy have evolved. Jain thinking overlaps with and arguably anticipates more recent scientific insights into the complexity of non-human species (Godfrey-Smith 2018), the centrality of minute beings to the survival of life on Earth (Fortey 1998) and the multi-layered nature of the universe (Bohm 2002). The emphasis on co-operation, whether between humans or between humanity and 'the rest' of nature is balanced by a rigorous belief in personal autonomy: the responsibility of each of us to think for ourselves and take ownership of our actions. There is no supreme being, but only a universal order the workings of which we are continuously attempting to understand.

From the environmentalist perspective, there seems to be little 'not to like' about the Jains. They can be seen as 'green pioneers', as repositories of an ancient wisdom that places humankind within the 'web of life' and asks us to live in harmony with nature. Jain commitment to animal welfare is famous the world over. The austere practices of ascetic men and women can easily be seen as exemplary as we seek to address the ill-effects of excessive consumption on the environment and our inner selves. This was broadly the view of the Jain tradition that I, as a sympathetic observer, held when I began researching my previous book, *Jainism and Environmental Philosophy* (Rankin 2018). However, I quickly became struck by the radical differences between Jain and 'green' thought and the incompatibility of Jainism with the western activist mode. While the latter is founded on a sense of urgency and moral certainty, the former is based on continual questioning, especially questioning of one's own thoughts, actions and underlying motives. Moreover, Jainism is far more than an eco-centric philosophy. Its ecological component is strongly embedded into the lives of 'civilian' men and women, the overwhelming majority of the Jain *Sangha* (spiritual community), but not the ascetics, warriors of the spirit engaged in a process of inner 'conquest' and detachment from ordinary or 'natural' sentiments and obligations. The Jain path is primarily concerned with the inner life, escape from the constraints imposed by the natural world (every devout Jain hopes to be an ascetic, if not now then in a future embodiment) and the liberation of the true self. Jains adhere to the 'ecological' principles of non-violence and care for the environment. Yet theirs is ultimately a 'religion of salvation' (Glasenapp 1999 [1925]) in which the only 'real' aspiration is the liberation of the self in its unfettered form. Enlightenment, identified with omniscience and immortality, means arrival at a state beyond the body and disentangled from the web of earthly life and the repetitive cycle of birth, death and rebirth. These are part of a less conscious state, an inferior mode of life that confers illusory attachments, defined as manifestations of karma.

This radical individualism is the essence of Jain spiritual practice. From a western standpoint, especially, it sits uneasily with a social philosophy of connectedness, co-operation and responsibility for others (human and non-human). From a Jain standpoint, by contrast, social and ecological responsibility are part of a longer-term process of detachment from material considerations. They stem from a recognition that each individual form of life has its own purpose, identity and viewpoint. Another apparent paradox is the comparative wealth of Jain communities, both within India and an extensive Diaspora. Many Jains enter professions such as medicine, law and education, but theirs is primarily a business community in which commercial success is highly valued. A strong 'work ethic' is inculcated at familial and communal levels, but wealth, like intelligence and knowledge, confers obligations. One of the reasons why materially successful businessmen and women are held in high esteem is the philanthropic tradition: the social pressure towards charitable giving is so strong as to be irresistible. Commercial success therefore benefits the whole community rather than only individuals and families. In India and the Diaspora, wealthy Jains also support educational institutions, hospitals, animal sanctuaries, human rights groups and environmental charities. Ascetic men and women are also sustained by business interests. Their austerities reflect a higher stage of spiritual development, but the intermediary stage represented by commercial activity plays a critically important role as well. The ethos of self-reliance and the pooling of communal resources is a logical extension of the principle of 'self-liberation' and responsibility for one's actions. It also reflects the status of Jains as a perpetual minority, having to preserve their culture, customs and thought against suppression or absorption. In so doing, they have not formed closed communities or attempted to shield themselves against ideas or influences from the outside world. On the contrary, the Jain ethos is one of adaptability, flexibility and learning from others, while preserving and building upon what is essential. In negotiating with the Hindu majority, Islamic rulers and Christian colonial powers, Jains have practised over many centuries the principles of 'soft power'.

Minority status has, for this small and distinctive population, been a source of economic strength and cultural consolidation. In response to the challenge of integrating and participating while preserving integrity, Jains have developed a tolerant yet intellectually rigorous worldview and achieved high levels of economic self-sufficiency. That intellectual rigour, combined with an acceptance of the validity of other points of view and a process of constant self-questioning, offers a good model for living successfully in a plural society and negotiating in a plural, interconnected world. Crucially, the principle of accepting and trying to understand alternative viewpoints extends to non-human perspectives. One of the reasons behind the concern with avoiding injury to even the tiniest life forms is that each of these beings is held to have its own unique viewpoint which is at the very least worthy of our consideration and respect.

In short, the Jain intellectual framework consists of 'soft' power combined with 'hard' rationalism, rigorous adherence to principle alongside a spirit of curiosity, questioning and experiment. It is this aspect of the Jain tradition, I argue below, that is of greatest relevance to environmental politics. We define that politics here as the attempt to rebalance humanity and the rest of nature, so that the priorities of the former are accommodated to the needs of the latter and the artificial division of 'human' and 'natural' is healed. Such healing need not, indeed should not, involve abandoning technological progress, although it requires us to re-evaluate the ways in which we use our technologies, indeed the way we apply our creative powers. Jain practice, applied to daily life, replaces an ethic of dominance with a principle of restraint, in which the merits and demerits of human activities are carefully evaluated for their impact on other living systems. By attempting to 'think like Jains' (Shah and Rankin 2017: pp. 19–39), we might reassess our relationship with the environment. In so doing, we reassess our purpose as human beings. The thought process of Jain philosophy goes beyond superficial 'radicalism' and examines the most pressing human and ecological problems from their roots.

Furthermore, Jain thinking and the way of life that accompanies it has evolved independently of western thought and experience, despite its acceptance of (and parallels with) western scientific logic (Mardia 2007; Mardia and Rankin 2013). For ecological consciousness to become genuinely 'global' it has to be plural as well. In particular, it needs to reflect the experience and accumulated cultural wisdom of non-western societies, sometimes referred to as the Global South. It is the peoples of Africa, Asia and South America who make up the overwhelming majority of the Earth's (human) population. They, and the plant and animal species that live alongside them, are on the front line of the current environmental crisis, much of which derives from uncritical application of the ideologies of 'progress' and 'growth' as ends in themselves. Many non-western nations, including India, are emerging as economic superpowers in their own right. Without their participation, all response to the environmental crisis is rendered unworkable and meaningless. The Jain worldview is therefore a useful case study in non-western ecological thought that has emerged from a literate, highly educated (in both western and traditional Indian senses) and in all respects 'developed' people.

The second chapter examines what is perhaps best defined as the spiritual science of the Jains, in particular their distinctive view of karma as an agent of interconnectedness. The 'web of life' in Jain doctrine is viewed at best with ambivalence, as a karmic entanglement to be eventually escaped. Yet part of that process of escape involves practical engagement with the external world, in particular conserving the environment and minimising harm. Jain ascetics, to whom some western green thinkers look for inspiration (Tobias 1991), lead admirable lives of restraint but are not directly relevant to environmental thought. Their path is one of almost complete withdrawal, abjuring nature as far as is humanly possible in the hope of an enlightenment that lies outside the material universe. It is to the majority of laymen and laywomen – the 'civilian' Jains – who are our subject of interest. Their way of life is based on accommodation with the natural world. Acting with care towards the environment is a spiritual practice that is built into daily lives, including those of the wealthy and materially successful. A distinctive conception of karma combines with a cosmology that emphasises the vastness, diversity and complexity of the universe. This engenders a long-term view in which actions are evaluated for their effects on future generations, including environmental impact. This approach bears some resemblance to the Seventh Generation Principle adopted by elders of the Haudenosaunee (Iroquois) people in North America, who represent another powerful current of non-western environmental thought (Rankin 2010, pp. 132–4). Jain philosophy is nonetheless concerned with transcendence of the world and ultimately liberation from the cycle of reincarnation and the attachments generated by karma. In this, it differs markedly from western environmental movements which are in general informed by secular, essentially worldly values accompanied at times by a romantic view of nature that embraces the 'web of life' as an unalloyed good.

Emphasis on asceticism and a one-sided view of Jainism as a 'green religion' are unhelpful, both from the perspective of understanding Jain thought and assessing its relevance to broader environmental concerns. Respect for nature is integral to the Jain way of life, but it is part of a much larger cosmology that is concerned with the journey of the individual soul or 'unit of life' towards freedom from karma. In Chapter 3, I propose that the aspect of Jain philosophy that is most relevant to our current concerns, at a global level, is the doctrine of multiple viewpoints or 'many-sidedness'. Many-sidedness, along with the doctrine of non-injury is the basis of the 'Careful Action' principle that governs daily life. Because every life form has its own 'viewpoint', it is worthy of protection. Many-sidedness is also a repudiation of ideological dogma and fanaticism, through a form of intellectual meditation in which our opinions and motives are called into question and tested. Many-sidedness enables its practitioners to step back, rethink and consider 'reality' from all possible angles. More radically, this doctrine without dogma reminds us of how much we do not to know and have yet to discover. This induces a state of humility in relation not only to fellow humans, and other human cultures, but the rest of nature. All activity is classified as 'karmic' because it impacts on other lives, but the karmic influence is diminished by considering the impact (in environmentalist terms, the ecological footprint) of all conscious actions. For generations of Jains, many-sidedness has served as an inoculation against fundamentalism, fanaticism and closed-mindedness. The inherently pluralist concept of 'multiple viewpoints' helps to explain the Jains' ability to integrate with the societies around them without compromising on their principles of non-violence and minimal environmental impact.

The final chapter explores the application of Jain principles to philanthropy, in which the concept of 'society' transcends the human sphere. The work of the humanitarian organisation Veerayatan is explored as an example of practical environmentalism without the tendency to romanticise 'pristine nature' that is all too characteristic of western greens. Veerayatan is run by female Jain ascetics who have not fully withdrawn from worldly concerns. Its projects, concentrated in the northern Indian state of Bihar, recognise that 'natural' forces such as floods and droughts have devastating impacts on subsistence farmers: nature is therefore a neutral, value-free force, by no means necessarily benign. At the same time, the effects of natural forces are often made worse by 'primitive' methods of agriculture as well as urbanisation and industrial processes. Veerayatan therefore views technological progress and material processes liberating and more in keeping with the principle of non-injury than poverty or apparent 'closeness to nature'. Such progress enables communities to withstand natural forces (modern homes are less easily destroyed, for example) and preserve their culture and way of life more efficiently. The appropriate use of modern tools and equipment lessens human impact on nature while practical environmental programmes such as reforestation mitigate the effects of flooding while improving the quality of life for humans and other species. Educational opportunities are also emphasised and viewed as enhancing local cultures as well as providing access to universal sources of knowledge. There is a characteristically Jain balance of continuity and change, which are seen as points on a continuum rather than polar opposites.

Veerayatan also supports small- and medium-sized enterprises (SMEs) as a means by which local communities preserve their economic and cultural independence. This stance is a development of Mahatma Gandhi's principle of self-sufficiency; the Mahatma was profoundly influenced by Jain ideas, including non-violence. Support for business is also an indication that Veerayatan (and other Jain-led philanthropic organisations) do not see any necessary contradiction between improving material conditions and improving human stewardship of the environment. The fanaticism of western 'eco-warriors' is absent from both their doctrines and their practical works. Their more nuanced position is partly derived from their origins in the Global South. In this context, material and technological progress offer liberation from poverty and drudgery and education (including western-style education) offers hope. Equally, the priorities of Jain philanthropists derive from a specific cultural approach to business whereby expansion and growth are not the most important priorities. Consolidation, the preservation of communal roots and the continuation of the culture of the firm are viewed as ultimately more important and as greater indications of 'success'. From a Jain perspective, there are natural limits to the size of a business. Commercial practice is closely partnered with philanthropy because charitable giving plays a central role in the strategy of any worthwhile business. Using the example of Vardhaman Gems of Jaipur, India, the final chapter outlines the many links between Jain business practice and the concept of 'minimal harm'. Awareness of an alternative business model is useful for western environmental campaigners. Their fully justified reaction against corporate excess and mechanistic forms of economics such as neo-liberalism often translates into a generalised hostility to business and an over-optimistic view of the state. With its emphasis on multiple viewpoints, Jain philosophy is a corrective to the 'hard' political ideologies that result in vengeful legislation and a culture of intolerance. For the ecological movement, such approaches offer only a dead end.

Any survey of environmental politics inevitably encounters problems of terminology. In drawing a distinction between 'environmentalism' and 'ecologism' or (the term I more often use) 'political ecology', I adhere closely to Andrew Dobson's definitions (Dobson 2007: pp. 10–27). Expressed succinctly, environmentalism is a reformist movement that emphasises incremental change while leaving the core values of politics and society for the most part intact. Ecologism is a more radical approach that can start with environmentalist goals but seeks a lasting change of cultural priorities, in particular a consciousness of 'the intrinsic value of the non-human environment' (Dobson 2007: p. 26). Deep Ecology, exemplified by the philosophical works of Arne Naess and the Bill Devall (e.g. Devall 1990; Naess 1989), is an important subset of ecologism, emphasising the spiritual aspects of engaging with nature as well as the value of cultural diversity. In their extensive writings, Naess and Devall draw strongly from their respective Norwegian and American *milieux*. Naess also refines Deep Ecology into a system and sensibility known as Ecosophy T; the 'T' refers to Tvergastein, a mountain hut in southern Norway's Hallingskarvet range where he spent extensive periods thinking and writing. Political ecology is my favoured term for the attempts to connect a broadly ecologist frame of reference with political activity and thinking. In Britain, for example, a think tank called the Campaign for Political Ecology was formed for this purpose in the 1990s, largely in response to the Green Party's embrace of far-left ideologies. I use the term 'green' either in the specific context of (western) Green parties or to denote a generalised (and again largely western) sensibility that underlies many forms of environmental campaigning.

'Environmental politics', in the context of this book, is a catch-all term for both environmentalist and ecologist activities. Some of these activities – and the ideas behind them – mirror or overlap with Jain perspectives while others differ markedly from them. The text includes many Jain terms which are explained both in their immediate settings and in the Glossary. In describing the small number of Jains who take ascetic vows, I refer variously to 'ascetics' 'Jain ascetics' or where relevant 'female' and 'male' ascetics. I consciously avoid the words 'monk' or 'nun' because of the misleading analogies with western Christianity to which these terms give rise. The term 'Jainism' is used for its familiarity in a western context, but where possible I have substituted 'Jain Dharma'. This is because, as I explain in more detail in Chapter 2, it is more authentic and broader, more accurate definition of this philosophy and the culture surrounding it. Finally, I use British spellings unless I am quoting from authors who write in American English.

Unlike Christians or Buddhists, for example, Jains do not seek to convert others to their faith tradition. Instead, they aim to find common ground with those who come from different philosophical backgrounds. In this way, they are able to exert subtle forms of influence on the societies around them in areas such as animal welfare, environmental conservation and the gradual move towards a plant-based (or predominantly plant-based) diet. Pluralism, the ability to listen (and not only to other human voices) and the reduction of unnecessary human impact on the natural world are central to Jain thought and of relevance to all forms of environmental politics. In an age when the expansion of human populations and the unchecked expansion of economic activity threatens the survival of many species of animals and plants, we can learn much from an Indic tradition that has long been aware of the connection between acceptance of multiple viewpoints and the maintenance of biodiversity. Seeing beyond the narrowly human perspective is probably the most important political challenge we face.

Bibliography

Bohm, D. (2002) *On Creativity* (ed. Lee Nichol), London and New York: Routledge.
Devall, B. (1990) *Simple in Means, Rich in Ends: Practicing Deep Ecology*, London: Green Print.
Dobson, A. (2007) *Green Political Thought*, 4th edn, London: Routledge.
Fortey, R. (1998) *Life: An Unauthorised Biography*, 2nd edn, London: Flamingo.
Glasenapp, H. von (1999) [1925] *Jainism: An Indian Religion of Salvation*, New Delhi: Motilal Banarsidass.
Godfrey-Smith, P. (2018) *Other Minds: The Octopus and the Evolution of Intelligent Life*, London: William Collins.
Mardia, K.V. (2007) *The Scientific Foundations of Jainism*, New Delhi: Motilal Banarsidass.
Mardia, K.V. and Rankin, A. (2013) *Living Jainism: An Ethical Science*, Winchester: Mantra Books.
Naess, A. (1989) *Ecology, Community and Lifestyle: Outline of an Ecosophy*, Cambridge: Cambridge University Press.

Rankin, A. (2010) *Many-Sided Wisdom: A New Politics of the Spirit*, Winchester: Mantra Books.

Rankin, A. (2018) *Jainism and Environmental Philosophy Karma and the Web of Life*, London: Routledge.

Shah, A. and Rankin, A. (2017) *Jainism and Ethical Finance: A Timeless Business Model*, London: Routledge.

Tobias, M. (1991) *Life Force: The World of Jainism*, Fremont, CA: Asian Humanities Press.

2 The Jain voice and political ecology

The colour of karma

In popular Jainism, a story is used to illustrate the specific significance of karma within this branch of Indian spirituality, expressed through the arcane theory of karmic colours. It is a useful, if perhaps unusual, starting point for a discussion of the complex interplay between the Jain philosophical tradition and political ecology. It is useful because it illustrates in clear language one of the esoteric teachings of the Jains, namely karmic colour theory. Simultaneously, it conveys a considerable amount of indirect information about Jain attitudes to the environment, in particular the relationship between humanity and the larger natural order, of which human beings are part. Of equal importance is the story's origin in the folk culture of the Jains, uniting their highly prized literate and scholarly tradition with lived experience and popular consciousness:

> Six men see a Jambu-tree [Syzygium, a member of the myrtle family], full of ripe fruit. They want to eat the fruit but the climbing-up is perilous to life. They reflect therefore as to how they can obtain possession of the jambus. The first proposes to hew down the tree from the root. The 2nd advises merely to cut down the boughs, the 3rd recommends to cut off only the branches, the 4th to cut off only the bunches [of ripe fruit]. The 5th wants only to pluck the [individual] fruit[s], the 6th at last says that one shall only gather and eat the fruit fallen to the ground. Here the first has a black, the 2nd a dark, the 3rd a grey, the 4th a fiery [or sometimes yellow], the 5th a lotus-pink, the 6th a white leshya [karmic colour].
> (Glasenapp 1991 [1942]: p. 48)

Before expanding on the significance of the karmic colourations cited above, and hence the overall meaning of the story, it is worth noting that the Jain theory of karma has a number of characteristics that mark it out from other Indic philosophical and religious traditions, including Hinduism and Buddhism. An examination of this theory necessarily involves imagery and concepts that might seem arcane, to secular western readers in particular. Their relevance to the connections (or lack thereof) between Jain thought and modern environmental politics might also be less than obvious. Yet the doctrine of karma is a key component of the value system of the Jains. It governs the decisions they make as individuals in their daily lives, and also defines their view of themselves in relation to the environment and their place within it. Moreover, beginning the discussion with karma is a useful exercise in 'thinking like a Jain' (Shah and Rankin 2017: pp. 19–39). Jain ideas can appear profoundly unfamiliar to the external observer, including the friendly commentator. Yet the most constructive response to this sense of strangeness is not the attempt to shoehorn them into more familiar western categories or equate them with political and social preoccupations with which on closer inspection they have little in common. The Jain system of logic is based on the questioning of all forms of dogmatic certainty. Therefore, it crosses as a matter of course the boundaries conventionally accepted by scholars and political campaigners alike. Crucially, Jain logic can also help us to arrive at a fuller understanding of the relationship between this branch of Indic thought and ecological consciousness. Its methods take us beyond immediate impressions that can mislead or even deceive.

Karma, for Jains, is defined as more than an abstract cosmic law of cause and effect or continuous cycle of actions and reactions. The theory of cause and effect, action and reaction, is central to a Jain's understanding of the workings of the universe. At the same time, it informs his or her attitude towards personal morality and ethics. Nevertheless, the concept of karma signifies far more than this. Karma is a substance more than a process, and it is visualised as consisting of particles of subtle, inert matter adhering to the jiva. The jiva is a life monad or unit of individual consciousness that every form of life contains, and which distinguishes living from inert matter (*ajiva*), which includes karmic particles. Even life forms considered elementary to untutored eyes possess jiva and the qualities of self-awareness associated with it. One of the most significant aspects of Jain thought is that all life is complex and mutually dependent. All living organisms play a distinctive role in the functioning of the earthly environment, and indeed the *Lokakasa* or 'inhabited universe'. The smaller the unit of life, the more powerful and ingenious its function is likely to be. This position arrived at through meditation, lived experience and scholarly insight is now increasingly echoed by the modern physical sciences. Oceanographers, for example, conduct detailed studies of the role of plankton in regulating sea temperatures and hence influencing the global climate (Fortey 1998: pp. 122–57). A species once dismissed by 'rational' humans as basic and unimportant is now recognised as essential to planetary equilibrium. In Jain communities, such conclusions have been taken for granted for more than two millennia. To them, a respectful and interdependent relationship with the natural world is not only 'right' from an ethical standpoint but also a form of rational self-interest. Individual life also has supreme value, so that everything that contains jiva or life essence has a common bond.

Every jiva, by the mere fact of coming into existence and acting spontaneously, comes into contact with karma, which is the product of *yoga* or activity of all kinds. Karmas, or 'karmic particles' (Mardia 2007; Mardia and Rankin 2013) act as obstacles to understanding the nature of the universe and the nature of the self, including the significance of jiva. Karmic particles are likened to specks of dust that darken a window, or to the sinews of a web that entraps. The latter comparison expresses ambivalence about the 'web of life' beloved of western ecological thinkers (e.g. Capra 1997). When western ecologists speak of life as an inter-connected web, they do so in celebratory terms. From their perspective, awareness of this 'web' is in itself emancipatory, with positive spiritual and political implications. From the Jain standpoint, the web represents entrapment by karmic particles. Awareness of the web is but a first stage towards emancipation. It is an understanding of the truth, or *vidya*, through penetrating the layers of karmic obscurity, which induce *avidya*: lack of understanding or ignorance. Understanding the role of karma is a step towards liberation from the cycle of birth, death and rebirth through the shedding of karmic particles.

In Jainism, as in western and indeed many human belief systems, knowledge is identified with light and ignorance with darkness. Knowledge as a means of overcoming karma is portrayed in terms of glimpsing light through a film of dust or dirt when it is gradually removed from the surface of a pane of glass. Knowledge, which is many-sided, is compared to a cut diamond reflecting light through its facets. Often, the jiva is equated with shining light and also with weightlessness. It is involvement with karma, especially through material attachments, that weighs the jiva down and keeps it locked in *samsara*, the repetitive cycle of existence and reincarnation. The goal of liberation is not ultimately arrived at through greater involvement with the temporal world, although there is a transitional phase when certain types of involvement, such as philanthropy or environmental conservation, are beneficial. Ultimately, *Moksha* or liberation involves transcending all attachment and activity in order to return to a state of pure being. Unlike most Hindus and Buddhists, Jains believe that individuality is retained after the jiva's liberation is achieved. Liberation, indeed, is the attainment of true, unfettered individuality. The journey towards release from karma follows a zigzag pattern of losses and gains, as opposed to a linear or progressive pathway. It can also be expected to last through many 'lifetimes' as the jiva passes through an assortment of animal, vegetable, mineral and human forms, until returning eventually to its point of origin. Unfettered by the presence of karmic particles, the jiva obtains omniscience. In its original or 'natural' form, its vibrations or movements (*yoga*) begin its encounter with karma and produce a cycle of further actions and reactions. There are many variants of *karma pudgala* (karmic matter), all of which either passively block out knowledge or actively distort our understanding of the universe. 'Our', in this context, refers not merely to humans but all living beings. Such influences affect the pattern of reincarnation and can either delay or accelerate spiritual progress.

Jains conceive of the karmic cycle as a natural process of physical and ethical evolution, but our 'destinies' are not wholly predetermined. The search for enlightenment involves a balancing act between karmic inheritance (the accumulated experience of multiple incarnations) and the conscious choices made by individuals in each discrete lifetime to stem the continuing 'influx' of karma and escape 'karmic bondage'. There are thus two parallel conceptions of individuality. First, there is the existing incarnated being of whatever form, the temporal self. Secondly, there is the life monad that passes through a range of evolutionary experiences and ethical choices, the cosmic self. Karmic accumulation may be seen as in some ways resembling genetic heritage. It confers some innate characteristics and points to a number of probable outcomes, without (in most cases) taking away the freedom to choose or the possibility of change. In seeking liberation, the more spiritually evolved beings are challenged to defy and defeat karmic destiny.

For human beings, who are viewed as the most spiritually evolved creatures in the hierarchy of life, the shedding of karma is achieved by good works, including care for the environment and acts of charity, not least towards other animal species. Yet because all activities are karmic, such good works generate their own form of positive karma or *punya*. This in turn overwhelms its negative counterpart, *papa*, and finally falls away to reveal an enlightened being. *Moksha*, the achievement of liberation, is associated with freedom from contact with the world of inert matter: jiva is finally separated from *ajiva*. In ethical terms, this means a process of emotional and intellectual withdrawal from material and even personal attachments. The goal is a state of equanimity (*Samyaktva*) and, as part of it, a position of non-interference and non-involvement. This goes beyond the reduced material consumption advocated by ecologists because it requires non-engagement with the rest of nature and as much independence from it as possible.

Philanthropy, conservation and the abandonment of materialism are therefore viewed as stages in a long-term process. So are achievements such as commercial success and accumulation of wealth, which in the political culture of western greens are often portrayed unfavourably and as evidence of contrasting values. The ascetic who is economically inactive and uninvolved in conventional social activity is regarded within Jain culture as the person who is closest to a liberated understanding of the true self. The liberated jiva, which has escaped from worldly definition, is a unit of life that is held to be all-knowing, permanent and motionless, hence the opposite of the life that circulates in *samsara*. *Moksha*, therefore, is simultaneously a return to the jiva's original form and the completion of its process of ethical evolution. Jains think in terms of cycles: the universe itself continuously expands and contracts in a series of forward and reverse motions lasting for aeons. The liberated jiva, unlike its original form, does not vibrate or engage with anything related to *karma pudgala*. Thus, the jiva's cycle is completed, but in a future so indefinite and remote that it is rarely directly considered in daily life. Yet every action (or non-action) is still in some sense a preparation for that distant goal.

Karma may be interpreted by practising Jains in either literal (or near-literal) terms or as an elaborate metaphorical explanation of the ethical dilemmas faced by humanity. Often, however, they see it in both ways, depending on the context in which it is considered or discussed. One of the richest images associated with karmic theory is that of *leshya*, or jivas as defined by the karmic colourations they acquire. These shades of colour represent gradations of spiritual development. The story of the Jambu tree at the start of this chapter is a popular parable explaining the system of leshya and relating them to the ethical choices open to human beings. It also tells us much about those Jain attitudes towards the natural world that superficially resemble, but on closer examination contrast profoundly with those of modern (and predominantly western) environmentalism. While the roots of what we might describe as 'karmic colour theory' are ancient, they should be seen as part of a living Indic system of values influencing millions of modern, highly educated men and women in India and an extensive Diaspora.

Leshya became clearly defined as a concept in the early centuries CE. In a literal sense, it gave new colour to a theory of existence that could seem pedantic and abstract to the lay practitioner. The jiva is held to enter each of its material embodiments reflecting one of six colours defining the karmic category by which it begins that phase of its journey. In acquiring colour, it passes from being 'free by nature from all distinctions perceptible by the senses' to 'receiv[ing] colour, taste, smell and touch' and so 'becom[ing] a defined type, which distinguishes it from other souls – although in a manner not recognisable by our senses' (Glasenapp 1991 [1942]: p. 47). 'Soul' in this context is the jiva as a unit of pure consciousness. The term has used with frequency in western translations of Jain texts, particularly before the latter half of the twentieth century. It is misleading when equated with, in particular, the conventionally held Christian concept of the soul, except in so far as it is perceived as eternal. In Jainism, the jiva is an animating principle. When it takes on a colour, this reflects and symbolises its temporary involvement with ethical concerns. The colourations divide individual jivas into the following karmic 'types':

1 *krishna*: black
2 *nila*: blue or 'dark'
3 *kāpota*: grey
4 *tejas*: red or 'fiery'
5 *padma*: lotus-pink
6 *shukla*: white

(Glasenapp 1991 [1942]: p. 48)

Some versions of this doctrine place a yellow (*pita*) hue in place of fiery red. The jiva itself is described as, for example, *shukla-leshya*. This state of 'whiteness' is 'characteristic of those on the highest rungs of the spiritual ladder' (Jaini 2001: pp. 114–15). The first three categories, black, blue/dark and grey, represent the 'heavy' or negative forms of karma (*papa*). In evolutionary terms, they are associated with beings held to have a low level of consciousness and capacity for moral choice. In ethical terms, they correspond to careless or harmful actions. Lotus-pink and white denote higher states of consciousness and hence a greater capacity to exercise both moral and rational judgement. They are also associated with the celestial beings that play an auxiliary, largely passive role in Jain cosmology and with 'light' or virtuous karma (*punya*). Red or yellow is an intermediate to higher state of consciousness associated with activity based on reason but not the highest form of reason. Humans and highly evolved ('five-sensed') animals 'show a very wide range of variation in moral development [and so] they may evidence any one of the six colours' (Jaini 2001: p. 114). Importantly, 'a being at its birth has in the beginning the leshya which it possessed at its death in the preceding existence' (Glasenapp 1991 [1942]: p. 49). In the course of a lifetime, it follows that the *leshya* 'can change' (*ibid.*). This change will be reflected in the next incarnation of the jiva and can go 'downwards' towards the dark or heavy colours or upwards towards the light. The hierarchy of colour is significant, darkness denoting ignorance (*avidya*) while white is equated with understanding or enlightenment (*vidya*), as it is the nearest earthly colour to the total clarity of perception of the liberated being. A human incarnation is more liable than any other to fluctuations between extremes of colour. This is precisely because of the advanced state of human intelligence and its potential deployment for creative or destructive ends.

The story of the reactions of six men to the Jambu tree and its fruit enables us to grasp the essentials of Jain environmental ethics. The most active suggestions are equated with the darkest and most negative karmas, the least active with the lightest and most positive colourations. This clearly implies an ethic of non-intervention. The most enlightened or 'ethical' choice is defined as gathering only the fruit that has fallen to the ground by natural means. Such a stance of non-action corresponds to the ascetic ideal of withdrawal from the world except where participation is absolutely necessary. Its practical limitations are obvious. Waiting for fruit to fall cannot satisfy immediate hunger or need for a nutritious diet. In any case, much of the fruit that has fallen is likely to be overripe and inedible. This attitude to nature is a way of opting out more than engagement. While it does not despoil the environment, like the first man's suggestion of 'hew[ing] down the tree from the root', it offers no solution to one of the basic human problems, namely satisfying hunger. Nor, crucially, does its stance of passivity allow for improvements to the environment, such as replanting or reforestation. The ascetic way of life might be the ideal but for most Jains it remains a remote objective because of its extreme impracticability.

It is therefore in the intermediate bands of colour, in particular the fiery red or yellow and the lotus-pink, where constructive human engagement with nature is to be found. This is the space where most Jains live and work. They have no intention of becoming ascetics, but they can exert a positive influence on the world around them. They do this by living as lightly as possible, showing respect and care by minimising harm to all life. At the same time, they can satisfy their essential personal needs and those of their families and communities. They may pursue commercial or professional success and enjoy its material fruits, provided that these are shared to make other lives better. Careful Action (*Irya-Samiti*) is central to the Jain ethos as an ascetic vow and, for the non-ascetic majority, a code of conduct based on personal restraint and non-violence. For ascetics, *Irya-Samiti* is a requirement that they have the barest minimum of involvement with the world, although in practice some ascetic orders, female ascetics in particular, become actively engaged in conservation and philanthropy. For the overwhelming majority of 'civilian' men and women, Careful Action means enlightened participation, through daily and often minute decisions about how to reduce harm to others, including minute forms of life. In Jain ethics, there is no clear distinction between the 'micro' and the 'macro', the large and small act. Actions themselves are defined by their underlying intention. Thoughts, positive and negative, count as forms of activity in themselves and have karmic repercussions for individuals and entire societies because of the more overt actions to which they give rise.

Many of the principles associated with Careful Action are familiar to environmental campaigners and professional conservationists alike. They correspond with already highly developed concepts of sustainability, the precautionary principle (Dobson 2007: pp. 59–60, 111; O'Riordan and Cameron 1994) and the ides of reducing humanity's 'ecological footprint' (Dobson 2007: pp. 71–2). The essential difference is that working with the grain of nature is an intermediate and not an ultimate goal for practising Jains. The sense of inter-connectedness shapes quotidian Jain morality, especially in relation to nature. However, being attuned to the natural world does not represent the highest state of consciousness, which is an uncluttered form of individuality, radically removed from worldly concerns or desires. Jainism is not a movement with political objectives. It is a spiritual system based on individual salvation through a combination of work and abstinence, involvement and withdrawal. Its doctrines are aimed at guiding the individual towards self-knowledge and understanding of the universe. Its social components, highly prized though they might be, should be viewed in the context of individual salvation and the extent to which they further this goal.

Jainism, in other words, is an otherworldly philosophy, concerned with transcendence of the material. Its principal concerns are with the cultivation of the inner self, which is identified as the source of knowledge. At the same time, and again confusingly for external observers, it is intensely pragmatic. Sustainable living is, as in political ecology and environmentalism, intimately connected with human survival. Yet there is also a strong emphasis on personal achievement and material success, whether for individual men and women or entire communities. Achievement, be it artistic, scientific, commercial or professional, is seen as a stage on the spiritual journey towards *Moksha*. Engagement with nature is not achieved through the conventional methods of *campaigning*, which involve adversarial or combative rhetoric and positions. Instead, a process of perpetual questioning and meditation gradually induces a change of consciousness. The viewpoint or *naya* of every life form is regarded as having its own validity, its unique story to tell. Each thing that contains the life essence of jiva is on the same quest for liberation from the cycle of birth, death and rebirth. In an atmosphere of political urgency, this worldview can easily appear conservative or obscurantist (Dundas 2002, 2004). Equally, the more extreme manifestations of asceticism may be conflated with extreme forms of political protest and dissent (Tobias 1991). Both approaches fail to capture the essence of Jain philosophy, in particular its inherent pluralism and acceptance of multiple viewpoints. We can find a distinctively Jain contribution to ecological thought in the perception of 'the environment' as a series of viewpoints that need to be considered in their own right and addressed on their own terms. Before exploring this doctrine of multiplicity at greater length or examining the differences between it and more 'orthodox' green perspectives, it is worth considering in more detail what it means to be a Jain.

Jain Dharma: a doctrine of equanimity

The Jains are a global community of up to ten million people, concentrated in India, their place of origin, but with Diaspora populations in eastern and southern Africa, North America, Europe and Australia. These populations retain close familial, cultural and economic ties. In India, there are approximately five million Jains, a population noted for its high literacy rate of 94.1% compared with a national average of 65.38% (Chakrawertti 2004). Significantly, they also have the nation's highest female literacy rate, 90.1% in comparison with the national average of 54.16% (Chakrawertti 2004). These statistics, drawn from the Indian national census of 2001, owe much to the community's high regard for education and the pursuit of knowledge by both men and women. Education, in this context, includes highly skilled apprenticeships such as those undertaken by the Jain jewellers of Jaipur, who are in many respects the pioneers of ethical business practice (Kothari 2004: pp. 48–50; Rankin 2018: pp. 1–6, 54–5). It is by no means limited to exclusively academic study, although that is highly prized, but encompasses lived experience, philosophical inquiry, critical reasoning and acts of compassion motivated by *Jiva Daya*: sympathy or identification with all sentient beings.

Ultimately, each individual is regarded as his or her own guru, but despite the existence of varying schools of Jain thought, there is a surprising degree of 'unity in diversity' and an absence of sectarian conflict. The lack of factionalism arises from one of the defining characteristics of Jain culture: a belief in the exercise of tolerance. The centrality of tolerance connected to the belief that reality is multi-faceted or comprised of multiple viewpoints. It is also closely linked to the emphasis on non-violence and the exercise of care towards the environment. The tolerance that is such a defining feature of the Jain way of life counterbalances the powerful ethical strictures that govern everyday life and the pressure exerted on individuals by expecting them to take responsibility for their actions and even their inmost thoughts. For Jains, dogmatic or inflexible ideologies are a symptom of *Ekant* (or *Ekantika*). *Ekant* is a form of 'one-sidedness' or doctrinaire certainty that stands in the way of greater knowledge, spiritual progress and the state of equanimity that needs to be cultivated to achieve both.

There is no distinction between the quest for scientific knowledge, especially knowledge of the natural world and the universe, and the quest for spiritual life. Both processes undermine and eventually break the cycle of *avidya*, the lack of perception that holds humanity (and individual human beings) back. It is *avidya* that leads to a distorted relationship between humankind and the environment, in which the former suffers from a delusion of superiority to the latter. This in turn leads to acts of harm to the environment that reduce the quality of human life and threaten human survival. *Avidya* is depicted in the form of a film of karmic particles obscuring the light of self-knowledge. As such, it induces a lack of scientific understanding of nature and an absence of spiritual affinity with it. These positions can be arrived at through 'pure' ignorance, distorting dogmas or a wilful refusal to evaluate evidence. In each case, the result is the same. The Jain theory of knowledge therefore involves a concept of humility in relation to the rest of nature and an acceptance of how little we know, individually or collectively. Knowledge is increased through a gradual, tentative process of questioning and testing ideas against each other.

India's Jains are one of the most successful business communities in the country and by some measures have the nation's highest per capita income (Chakrawertti 2004). Overwhelmingly a population of city-dwellers, they have a strong numerical presence and cultural influence in professions such as medicine and law. This pattern is replicated in the Diaspora, where the Jain story is one of consistently high educational attainment followed by high achievement in business, finance and the professions. Medicine and law are, again, frequent choices, along with optometry and the jewellery industry. In Antwerp, for example, one of the leading global centres of the diamond polishing trade, Jains of Guajarati and Rajasthani heritage have achieved great prominence and exercise a considerable degree of 'soft power' (Hofmeester 2013; Lum 2014; Meena Peters 2014; Rankin 2018: pp. 54–5, 59, 63; Shah and Rankin 2017: pp. 16, 51, 85, 116). Diaspora Jains have mostly achieved a judicious blend of integration with their host communities and preserving, often with great care, their underlying cultural values and belief system. The practice of engagement with the world, through the pursuit of educational attainment along with commercial and professional success, is not seen as contradicting the ultimate goal of material renunciation. There are two main reasons for this. First, worldly knowledge is regarded as a stage in spiritual growth. Secondly, Jain logic is based on the concept of both/and in place of either/or: differing ideas are viewed as points on a continuum, not irreconcilable opposites.

The word 'Jainism' is used for convenience in this discussion because it is the term most widely understood outside the Jain community in India and overseas. Many Jains have ambivalent feelings about the word, with a few rejecting it altogether. This is because they consider the concept of an '-ism', implying a tightly defined, compartmentalised ideology, as an insufficient or misleading reduction of their doctrines and way of life. The term 'Jainism' can also imply a closed or insulated system of values, separated from the main currents of Indic culture. In itself, this idea makes many Jains uncomfortable. While they do insulate certain aspects of their faith tradition and go to some lengths to preserve their identity, they are also keen to emphasise continuities with the wider culture around them. Instead of asserting their separation from the Hindu majority, they often incorporate Hindu deities and rituals into their worship or meditation, regarding them as important parts of their heritage. Equally, they emphasise the compatibility of their worldview with democratic institutions and the pluralist, secular values underpinning them. To most Jains, the theories and practices called 'Jainism' by outsiders are components of Dharma, at once a cosmology, an ethical system and a practical code for daily life. The phrase 'Jain Dharma' is, therefore, frequently (but not invariably) used below as a replacement for the more familiar '-ism', in particular when exploring its philosophical tenets and their applicability to environmental thought.

An approximate translation of the Sanskrit word Dharma is 'cosmic order', although more literally it means 'that which holds the world [and the universe] together'. This concept encompasses but also transcends the mainstream western understanding of 'philosophy', 'religion', 'faith' or 'ethos'. For practising Jains, these are all intersecting parts of a larger 'whole'. In the same way, sitting meditation, reverence for the *Tirthankaras* (the 'ford-makers' or 'path-finders' who serve as guides to enlightenment), charitable giving (including support for the *Panjrapoor*: animal hospitals or sanctuaries for which Jains in India are noted) and the conscious avoidance of harm to all species are aspects of Dharmic practice that reinforce each other and reduce karmic influences. Jain Dharma is also more than a series of doctrines lived out by loyal adherents with varying degrees of success. As well as providing rules for daily living and a cosmological framework, Dharma is a cultural and aesthetic sensibility, reflected in the serene images of *Tirthankaras* and representations of the universe in painting and sculpture. These detached, all-knowing 'peaceful liberators' (Pal 1994), free of all mundane concerns, including dependence on the environment. At the heart of this calm disposition is a sense of perspective based on long-term thinking coupled with an understanding of the universe as vast, eternal and cyclical. Kanti Mardia, the Jain mathematician, refers to this sensibility as 'Jainness'. This idea combines the intuitive, eco-centric components of the Dharma with its humanistic, rational and transcendent aspects. At the same time, it expresses a human dimension:

> Jainness is concerned with reconciling continuity and change, possibility and limitation. It offers the possibility for spiritual liberation through self-knowledge, while accepting the mental and physical limits of the human form. ... Rather than seeking to impose its own truths, [Jain Dharma] asks us to look inside ourselves, find our own truth and continually question it.
> (Mardia and Rankin 2013: p. 26)

We have already noted the centrality of the individual in Jain doctrine. This distinguishing characteristic has undoubtedly contributed to Jain success, including commercial and financial acumen, because of the pressure it exerts on each person to fulfil his or her potential. However, this individualism is very different from that of classical liberal thought, based on the autonomous actions of a time-limited entity. Dharmic individualism, as we might call it, is based on a 'self' that progresses (or regresses) from one embodiment to another until the point where it achieves knowledge of its true nature, detached from the karmic 'web of life'. When that happens, individuality remains but as a unit of pure consciousness without the physical limitations associated with the cycle of existence known as *samsara*: birth, death and rebirth. In the most traditional versions of Jain cosmology, the enlightened being or *siddha* (liberated jiva) rises to the top of the universe, a process that symbolises the highest point of self-realisation and freedom from the weight imposed by karmic matter. While the drive for personal and communal success is very much present in Jain society, it is seen from the perspective of multiple incarnations and the expansion and contraction of the 'inhabited universe' over millions of years. It is not therefore seen as an ultimate goal, but merely a step on a pathway. Financial or professional success, along with greater knowledge, confers extra responsibilities to act with care and 'give back' the fruits of that success.

Jain doctrines have been transmitted from pre-literate antiquity to the present through a series of *Tirthankaras*, or 'ford-makers' who have liberated themselves from the cycle of *samsara*. *Samsara* itself is defined both as the cosmic cycle of existence and the processes of everyday life. Liberation from *samsara* involves attaining the status of Jina: one who has overcome inner passions and to whom material and temporal (including environmental) considerations no longer have value. The attainment of this level of self-knowledge is a springboard towards becoming a *siddha* or 'liberated soul': the life essence in a refined form. A *Tirthankara*, in turn, is a special type of *siddha* whose purpose is to guide others by explicit or implicit example. *Samsara* is depicted as a *tirtha* or ford to be traversed and the *Tirthankaras* act as pathfinders or points of light on that journey. Many Jains use them as the focus for meditation, sometimes through images or statues, in other cases through techniques of visualisation. This is not the same as worshipping deities. The ford-makers are not regarded as 'gods' but as inspirational examples and role models. The devotee uses them as points of reference while searching within him or herself for practical answers to problems in living, or sometimes considering larger spiritual questions.

Most of the 24 *Tirthankaras* are prehistoric figures whose lives are barely remembered except through hagiographic fragments. The mythical lawgiver Rishava (also known as Rishabha or Rikhava) is the first *Tirthankara* and he is also recognised by some Hindus as aspect of the deity Shiva, an example of the complex interplay between the two Dharmic traditions. Parshva, the 23rd *Tirthankara*, is the first whose historical existence can be fairly reliably verified. He is believed to have lived at some time between 872 and 772 BCE (Zimmer 1969: pp. 181–205). The discernible 'modern' philosophy of Jain Dharma emerged in a coherent form at the time of the 24th *Tirthankara*, Vardhamana Mahavira (559–557 BCE). Mahavira's name means 'Great Hero'. Heroism, in this context, means spiritual victory (the Jina as 'Conqueror') and self-realisation through non-violence and abjuring conventional power or wealth. Mahavira was a contemporary of Gautama Buddha, but he favoured more radical austerities, the origins of the vows undertaken by ascetic Jains today. Asceticism is balanced by the pragmatic, socially engaged and environmentally aware practices of lay men and women, based on a modified version of the vows. Mahavira plays a prominent role in Jain popular culture, again not as a divine figure but as a source of inspiration and practical wisdom. Many Jain households and businesses have statues or images of Mahavira and Parshva, although others prefer to view them in their more abstract forms as fully realised beings: unfettered jiva or pure consciousness.

The traditional Jain greeting *Jai Jinendra* can be translated as 'Hail to the Conquerors' or 'Honour to the Supreme Jinas'. On first reading, the references to conquest have connotations of militarism and dominance, but in fact they signify the reverse. For the Jain, conquest is an internal process of self-discipline to overcome negative aspects of the self that are associated with attachment, desire and *kasaya* or 'passions'. These destructive emotions range from material acquisitiveness as an end in itself to fanaticism and doctrinaire certainty, leading to cruel and harmful acts and hateful thoughts. Concepts such as evil and 'sin' do not exist in Jain Dharma. Instead, the emphasis is placed on positive and negative developments, that is to say behaviours or thoughts conducive to enlightenment and behaviours or thoughts that perpetuate ignorance and 'darkness'. Individual conscience and individual consciousness are central to Jain theory and practice. Enlightenment is arrived at through the gradual reduction and shedding of karma, a process that corresponds with the non-violent activity, the dispersal of needless possessions and avoidance (wherever possible) of actions that impact adversely on the environment. Inner discipline is required for these tasks, which is why observant Jains undertake to follow five vows (*Vrata* or *Vratas*) attributed to Mahvira. These are rooted in a sense of personal and social (including ecological) responsibility but they are also a response to karmic influence and lead to eventual release from all material commitments, at which point the jiva returns to its essence. The Five Vows are generally listed in the following order:

Ahimsa: Non-violence, non-injury, respect for all life
Satya: Truthfulness, honesty, personal integrity

Asteya: 'Non-stealing', avoidance of theft, taking what is not given or offered, exploitation of human beings, animals or natural resources

Aparigraha: Non-possessiveness, avoidance of wasteful consumption, unnecessary accumulation of possessions or viewing possessions as ends in themselves

Brahmacharya: Chastity, avoidance of promiscuity or exploitative relationships

Ascetics undertake the *Mahavratas* (Greater Vows) whereas lay men and women practice the *Anuvratas* (Lesser Vows). In the case of Brahmacharya, for example, ascetics interpret the vow as requiring celibacy, whereas for lay Jains it means fidelity in relationships and mutual respect between partners. The wider significance of the vow is not so much connected with sexual behaviour, or even intimate relationships, but responsibility towards others. The *Vratas* do not serve as discrete injunctions but intersect like circles in a Venn Diagram. The overarching theme is the reduction of harm (*himsa*), whether towards oneself, the environment or fellow human beings. For each individual, the level of personal and social obligation increases with their levels of material (and hence karmic) accumulation. Possessions are not viewed in simple terms as a burden, but as an opportunity for personal development and positive social or ecological acts.

The first principle is Ahimsa, the theory and practice of non-violence, or in a more literal sense, non-injury and the absence of harm. Ahimsa is a thread that connects all aspects of Jain Dharma. For those who live by the *Anuvratas*, it means actively seeking to reduce the levels of harm enacted by each person and by humankind collectively. In practical terms, this means refraining from or minimising actions that damage or threaten other forms of life. This is why Jains maintain a strict vegetarian diet, a hallmark of their culture of which they are proud. Plants, like all living things, contain jiva, but do not (according to Jain theory) have as many 'senses' or level of consciousness as animals (Mardia and Rankin 2013: pp. 84–94). At the same time, they sustain human life and health, and so are treated with respect and care. For the lay man or woman, the practice of Ahimsa combines Careful Action in everyday life with positive measures to minimise or alleviate harm. Charitable donations and voluntary work therefore qualify as Ahimsa, as does exercising responsibility as a professional or business owner, in particular practising transparency and avoiding deception or exploitation. In these contexts,

Ahimsa intersects with Aparigraha (charitable giving, divestment of unwanted or unnecessary possessions) and *Asteya* (avoidance of theft, dishonesty and exploitation). The Vratas are mutually dependent, reflecting the Jain view of life as a whole. Non-injury and active alleviation of harm generates *punya*, the positive, 'light' form of karma and increases self-awareness. For ascetics, by contrast, the *Mahavrata* of avoiding harm to others requires progressive withdrawal from all activities that *could* have a negative impact on the self and others. This does not always preclude charitable works, including education, but the most 'advanced' ascetics reduce to the barest essentials their involvement with humanity and the rest of nature. Their practice of non-action, or minimal action, leads to the more rapid shedding of karma.

It should be emphasised here that the overwhelming majority of Jains seek to live in accordance with the *Anuvratas* and in so doing engage actively and constructively with the world as they find it. At the same time, they are working towards personal salvation. Social (or environmental) action is undertaken with that transcendent aim in mind. The ethical code of the *Anuvratas* is at one level strict and uncompromising, at another pragmatic and flexible. From a western perspective, it can seem rigorous and sometimes uncompromising in the connections it makes between apparently small acts of individual carelessness and larger collective acts of harm. Walking without thought across an area of grass and not following the paved path is seen as being as much a form of *himsa* as air pollution caused by traffic and bad public transport or the poisoning of rivers with industrial waste. The first example is objectively far less serious than the second and third, although the lasting impact on microorganisms, insects and other small forms of life is not dismissed as trivial by Jain ethics. Each living being has its own valid viewpoint, the ecological significance of which is fully acknowledged even if its value is scarcely perceptible to most human beings. Attitude and intention link all forms of *himsa*, whether large-scale or apparently unimportant. Collective carelessness begins with individual acts, in the same way as negative or violent thoughts produce negative or violent outcomes. However, the Lesser Vows are flexible insofar as they recognise differences in personal circumstances. The less wealthy might not have the same range of choices at their disposal as their more privileged contemporaries. It follows that they are likely to be less able to avoid small acts of harm and have less time to devote to meditation or study. Equally, wealth and education bring added responsibilities to act with care. Each person is at a different stage on the spiritual journey, influenced by ability and circumstance as well as the choices he or she makes.

Jain Dharma is a non-theistic system. Deities can be recognised and revered although they function more as cultural reference points linking Jains to their surrounding communities. In India, for example, they often participate in local or regional Hindu festivals honouring popular gods and goddesses (Jain 1993). While it is the *Tirthankaras* who have ultimate significance for Jains, the devotion to local Hindu deities in India and the frequent sharing of facilities with Hindus in the Diaspora are both important examples of Jain pluralism in action. No creator deity or First Cause is acknowledged, and the universe is viewed as eternal, constantly pulsating and reinventing itself as it passes through regressive and progressive half-cycles (*Avasarpini* and *Utsarpini*). Jain Dharma emphasises neither permanence nor impermanence, for both are of equal significance. In the 'inhabited universe' there is a permanent creative tension between continuity and change, the 'persisting unity (*ekatva*) of the substance [or object, person, animal, plant, etc.] and the transient multiplicity of its modes' (Jaini 2001: pp. 90–1).

This cosmic struggle also takes place within each individual. The jiva is permanent, whereas its current incarnation is impermanent and transient, eventually fading and dying. Taken together, the absence of a creator god and the tension between continuity and change have two main effects on the Jain outlook. First, a high level of responsibility is placed on individuals and communities alike to interpret and apply the *Vratas*. With that comes freedom to discover, think and create, which is why scientific experiment and innovation have always been valued highly in Jain society, along with philosophical speculation and artistic creativity. All of these, in different ways, illuminate aspects of a multi-faceted universe. The second effect is to inculcate a sense of perspective based on the vastness of the universe and the perishable quality of all material things. What appears to be solid is in fact impermanent and so it follows that abandoning attachment to the material is to acknowledge an underlying reality.

As well as balancing continuity and change, Jain Dharma aims to unite intuition with reason. The rational and intuitive aspects of human thought are of equal value, in different ways, and exits in a continuous state of interplay. A 'eureka moment' of scientific breakthrough would be recognised as a point at which these two principles meet. The starting point for the Jain view of nature is an intuitive awareness that everything within it is 'alive' coupled with a sense of wonder. This sensibility remains an important defining feature of 'Jainness', but it quickly gives way to a rational awareness that all life is mutually dependent, and our actions impact on the lives of others.

Jiva Daya is defined as sympathy with all sentient life, which is why animal welfare is accorded a status equivalent to human welfare in Jain morality. One of the guiding precepts of that morality and today effectively a motto for Jains is a Sanskrit verse from the *Tattvartha (That Which Is) Sutra* of Umasvati, composed between the second and fifth centuries CE: *Parasparopagraho Jivanam*. This phrase can be translated in two ways: the secular and philosophical 'All life is bound together by support and interdependence' and more explicitly religious and devotional 'Souls render service to each other', souls in this context being the eternal, inalterable aspect of the jiva.

Both translations of this phrase are of equivalent value. The difference is one of emphasis: the first translation appeals to the rational or scientific mind, the second to the intuitive or artistic sensibility. For many Jains, *Parasparopagraho Jivanam* has become a popular aphorism, signifying the belief in non-injury coupled with a pluralist method of problem solving. This form of pluralism extends beyond humankind, so that human activity has to take account of the interests or 'viewpoints' of other species and ecosystems. As a motto for all Jains, it is written in *Devanagari* script at the base of the emblem adopted in 1994 to mark the 2,500th anniversary of the Nirvana of Mahavira. This date has special significance for Jains of all schools of thought because Mahavira is held to be the final Tirthankara for the current half-cycle of the universe. As a principle of interdependence, *Parasparopagraho Jivanam* operates at the levels of friendship and family ties, the ethical conduct of business and interactions with the natural world. Attitudes towards the environment are thereby integrated with beliefs and practices affecting a wide range of activities, from education to household management and financial transactions (Shah and Rankin 2017). 'The environment' itself is not defined so specifically as it has tended to be in the West since the industrial era especially. It encompasses human society as much as the natural world, thus eroding the artificial divisions between human and animal realms. This is why the Jain definition of animal welfare as in effect equal to human welfare can seem so radical and far-reaching to non-Jains. In fact, it stems from recognition of the shared characteristics of all sentient forms of life.

Jain Dharma is holistic in character, uniting secular philosophy with religious devotion, scientific reasoning with creative intuition, scholarly knowledge with a sense of enchantment. Engaged with the world, yet seeking liberation from material constraint, it challenges the western obsession with compartmentalising ideas or areas of knowledge. The divisions between the arts and sciences, theory and lived experience seem strange and superfluous from a Jain perspective. Our certainties and processes of reasoning are called into question by this approach, which identifies spiritual progress as the ability to ask questions and not know the answers.

The challenge to western thought posed by Jain philosophy can be seen as implicitly but not overtly political in character. It is not a challenge in the most obvious sense, as according to Jain logic western-derived ideas are a series of *naya*, or viewpoints worthy of respect. In any case, Jains are willing and successful participants in many western countries, contributing to the economy and society while accepting the values of the majority culture alongside their own. They tend to look for parallels between the two worldviews and avoid too much emphasis on areas of difference or disagreement. Aparigraha and the concept of Careful Action resemble in some respects the concerns of the green political movement as it has developed in the Global North. However, green politics arose in response to industrialism and the culture of mass consumption, whereas Jain Dharma evolved in a highly stratified, predominantly rural or mercantile culture now embracing urbanisation and consumerism. Common ground can certainly be reached between the key tenets of western political ecology and Jain culture's profound concern with non-injury to the environment and preserving the equilibrium between humanity and the natural world. Yet the partisan ideology of green politics and the intricate system of Jain Dharma evolved in very different conditions and serve distinct and at times incompatible purposes. Before we make any attempt to fuse a number of highly specific Jain and green values, the differences between the two doctrines deserve our consideration.

The 'greening' of Jain Dharma?

The vow of non-possessiveness, Aparigraha, is an apparent point of connection between Jain philosophy and green political thought. It asks practitioners to view the accumulation of wealth, property and other possessions less as an achievement in itself and more as a means to two ends. The first of these is social, by which wealth is used for practical purposes to improve the lives of others. The second is, in the broadest sense, spiritual. Rejection of materialistic values and the acquisitive mentality is a move towards the state of equanimity and detachment associated with enlightened knowledge and escape from the influence of karma.

Aparigraha lessens consumption and so in turn diminishes the harmful effects on the environment of human activities. The vow is in one sense 'passive' or inner directed, favouring renunciation and contraction. At another level, it can be seen as 'active' or expansive, enjoining practitioners to be active citizens and charitable donors. There are also, as with the other vows, two forms of Aparigraha. Ascetics take the idea of renunciation as literally as possible, subsisting with a bare minimum of possessions. Lay Jains, by contrast, seek to reduce their dependence on material things as far as is practically possible given their family ties and business and professional commitments. The practice of Aparigraha requires not only the positive social use of income and wealth, but also the gradual divestment of possessions as they become unnecessary to maintain a reasonable quality of life. Methods of divestment include donating income to charities (human or animal), schools and universities, hospitals and environmental projects, or simply giving to individuals and families in need. It can also mean playing a more active role in setting up charitable foundations and contributing to the cultural life of fellow Jains or the wider community. Equally, it can involve recycling money into a family business so that younger members of the extended family assume greater responsibility and employees also benefit. No strict distinction is made between 'earned' and 'inherited' income. This is partly because of the cultural emphasis placed on collective family endeavour, but also because who actually earned the money is viewed as less important than whether the money has been acquired by ethical means and is applied to ethical ends. Needless to say, the ultimate sources of that wealth have to be rooted in the ethics of Ahimsa and involve minimal harm to living beings.

The accumulation of wealth brings with it heavy moral responsibilities. It confers the power to do greater good but equally increases the potential to inflict harm, through exploitative behaviour or overconsumption of resources. In this sense, wealth is analogous to knowledge, the source of creative and destructive powers alike. Knowledge itself is viewed as a process of accumulation, the benefits of which should be shared with others. The vow of non-possessiveness is also associated with seniority. Age provides the opportunity to withdraw from the process of accumulating possessions and at the same time reduces material needs. The responsibilities of work and family are transferred to a younger generation, while the old share the benefits of their wealth and understanding of life. With seniority comes abdication from the role of householder and provider and with it a sense of calm that can lead to *Samyaktva*, the benign objectivity that is a precursor to enlightenment. Aparigraha also influences the conduct and structure of Jain businesses, keeping them anchored in their community. Successful business practice is not necessarily equated with unlimited expansion. On the contrary, the Jain business ethos is defined by caution and an awareness of when to stop expanding, or even contract. Far from being politically correct 'add-ons' or window dressing, values are viewed as integral to commercial practice and are part of the definition of 'success'. The idea of having to choose between values and efficiency is culturally alien, indeed disconcerting, from the many-sided standpoint of Jain Dharma.

Jain attitudes towards business ethics and practices will be examined in greater detail in Chapter 4. For now, it is worth noting that they take account of 'limits to growth' (Meadows and Meadows 1972; Meadows and Randers 2004), perceiving unlimited economic expansion as a threat to the underlying ethos of a commercial enterprise and its role in the community. Unlimited expansion is also seen as a potential cause of dangerous economic and environmental imbalances. Applying limits to growth also preserves the culture of a company and the distinctively Jain values it seeks to promote, which would be lost or diluted in a corporate bureaucracy. These values, governed by the principle of Careful Action and the vow of non-possessiveness, are as we have seen compatible in many ways with the objectives of political ecology. Their emphasis is nonetheless subtly different, because it is based on personal conscience and voluntary actions in place of collective consciousness or legislative action. Its ultimate goal is detachment and withdrawal, as opposed to engagement in the world. Reconnection with nature is but a step towards transcending the world of matter.

For Jains, there is no appreciable difference between cultural and environmental conservation. Both of these principles are based the value of continuity (balanced by rational, incremental change) and the need for co-operation with fellow-humans and (as far as possible) all living entities that contain jiva. As such, Jain environmental consciousness is not universal but specific, an aspect of the sensibility described as 'Jainness', through which an ancient culture is preserved in a modern form and retains its vitality. We have noted above that no distinction is made between 'society' (human) and 'the environment' (essential to human life): one is seen as an extension of the other. The stance of inner-directedness, coupled with a belief in private property and personal initiative can make Jains seem apolitical, quietist or 'small-c' conservative. Such conclusions simplify a more complex picture. While Jain Dharma is not a proselytising faith, its ethical doctrines present a quietly radical challenge to the prevailing model of economic growth, whether the engine of growth is the private sector or the state. For a sense of common purpose to emerge between Jain doctrines and the aims of environmental politics, the fine-tuned differences of perspective should be understood and appreciated. The table below (Table 2.1) aims to tease out these distinctions as well as pointing to areas of overlap or potential common ground.

These classifications share the flaws of most forms of taxonomy, for although they capture the most significant details, they do not express anything like the whole picture. They therefore require a number of important caveats. The social conservatism of the Jains, for example, is evident in their emphasis on the traditional extended family and their powerful sense of a cultural identity that should be preserved and sometimes shielded from outside influences. Jains believe in continuity in the midst of radical external change. This includes continuity from father to son or mother to daughter, the preservation of the structures and emotional ties that bind a community together and protecting the

Table 2.1 Jain Dharma and Environmental Politics: Contrasting and Overlapping Characteristics

Jain Dharma	Environmental Politics
Careful Action (Irya-Samiti)	'Precautionary Principle'
Ahimsa/non-injury/Punya (positive karmic activity)	Non-violent direct action (NVDA)
Organic evolution in India	Emerged (in Global North) in reaction to industrial society
Socially conservative	'Progressive' or radical on social issues
Indirectly political	Partisan and campaigning
Transcendent: seeks liberation from Ajiva or material world	Seeks liberation through integrating with nature
Conservation of nature as a step towards self-realisation	Conservation of nature as end in itself (or as part of broader programme of social transformation)
Centred on the individual	Collectivist
Pro-science: supportive of technological solutions as long as they minimise harm	Ambivalent and conflicted over role of science and technology
Pro-business	Ambivalent or hostile towards business
Philosophical system and cultural sensibility	Political ideology
Specific (rooted in Indic culture)	Universalist (rooted in western culture)
Does not seek converts	Actively proselytising
Voluntary action	Collective action
Spiritual emphasis	Material emphasis

ethos of a commercial enterprise even at the cost of opportunities to expand (Rankin 2018: pp. 54–63; Shah and Rankin 2017: pp. 74–93). Such 'small-c' conservatism is counterbalanced by an emphasis on the education and economic independence of women. This is derived from a historic belief in equality of the sexes. Mahavira, in contrast to Gautama Buddha, had more female than male disciples and there have always been more female than male ascetics in the Jain *sangha* (faith community). Underpinning this commitment is a respect for both women and men as individuals with the right – combined with personal and social obligation – to make use of their full range of abilities. Environmental politics, classified above (and by most other commentators) as socially radical, also contains a conservative sub-theme in the form of the precautionary principle, in many respects a secular version of *Irya-Samiti*:

> The implied impossibility of knowing enough is crucial to the green suggestion that we adopt a cautious approach to the environment. If we cannot know the outcome of an intervention in the environment but suspect that it might be dangerous, then we are best advised, from a green point of view, not to intervene at all. ... In this respect, green politics opposes drawing-board social design and thus falls into the realm of what is generally considered to be conservative politics.
>
> (Dobson 2007: pp. 59–60)

Similarly, the active principle of *punya*, or 'positive karma', can overlap with the practice of Nonviolent Direct Action (NVDA) associated with environmental activism. More often than political protest, *punya* takes the form of constructive activities in support of disadvantaged communities or animal welfare. The work by the female ascetics of the Veerayatan movement in rural Bihar serves a case study in *punya*. By combining education and training with environmental conservation, it makes the same link as green politics between social and environmental justice. The importance attached to commercial activities by Jains is underscored by a rigorous attitude to business ethics based on Ahimsa and Aparigraha. As far as there can be said to be such a thing as 'Jain economics', it favours local production for local need, provided by small and medium sized enterprises or co-operatives. *Swadeshi*, Mahatma Gandhi's economic model for India, derived many of its ideas from Jain ethics, including a belief in sustainable or intermediate technology and economic self-sufficiency (see Chapter 4). In the same way, Gandhi was inspired by the Jain merchant-scholar Shrimad

Rajchandra to re-evaluate the principle of Ahimsa within his own Hindu tradition. A pioneer of Nonviolent Direct Action, his campaign of *satyagraha* ('truth-struggle') against British colonial rule was influenced by Rajchandra's thinking (Shah and Rankin 2017: p. 84).

Jain culture goes far beyond Gandhian principles in its embrace of technological and scientific innovation. Whereas the Mahatma and his supporters viewed technology with caution, fearing its dehumanising effects, the Jain ethos favours experimentation and embraces modernity while in an apparent paradox preserving ancient traditions. This open-mindedness is a reflection of the holistic form of Jain logic, favouring multiple over binary choice, both/and over either/or. From this standpoint, tradition and modernity are far from incompatible, but strengthen and reinforce each other. There is also a powerful Jain tradition of scholarship, rational inquiry and (under the influence of 'many-sided' logic) a process of continuous questioning and revision of ideas in the light of evidence and experience. This fosters a belief in the socially improving power of science and technology. Like support for business, the pro-technology stance surprises non-Jains, who are often captivated by images of austerity and asceticism (Tobias 1991). Yet many Jains are highly aware of the liberating possibilities of technology as well as its contributions to human understanding, which is part of the spiritual journey. It is the way in which scientific and technological understanding is applied, and the intentions behind the application, that determine whether they serve as civilising or destructive forces. Overall, Jains take an optimistic view of scientific progress, not dissimilar to that of the European Enlightenment, except that they favour a 'holistic' or 'integrative' over a 'mechanistic' approach. In particular, they emphasise the appropriate and ecologically sensitive use of technology to minimise harm and improve lives.

Environmental politics, by contrast, is often profoundly sceptical about scientific and technological change. The green movement has emerged as a critique of the prevailing assumptions of the industrialised West that seem to confuse quantity with quality, assuming that resources are boundless, as opposed to finite, and that linear progress, defined largely by technology, is both inevitable and inherently desirable. The ecological sensibility fused with the New Left critique of capitalism in the political and cultural ferment of the late twentieth century to create the green movement in its most visible and organised forms (Benton 1993, 1996, Bookchin 1972, 1982; Weston 1986). Ecological consciousness is frequently expressed through evoking a green sensibility. Like 'Jainness', it is difficult to position accurately on the left-right spectrum of conventional politics.

One of the principal features of the green sensibility is a perception that the emancipatory aspects of the Enlightenment, including personal and intellectual freedom, have come at a substantial ecological and spiritual cost. Enlightenment values have, when viewed from this angle, brought with them a narrow and instrumental view of progress. With that has come a restrictive definition of the individual as *homo economicus* ('economic man'), shorn of social or cultural context, and the alienation of humanity from the natural world. The prevailing Jain perspective, by contrast, dwells less on the negative effects of technological progress and more on the possibilities offered by careful use of technology, such as freedom from the physical drudgery, rural isolation and disease. Rather than adopting a romanticised view of poverty or rural isolation, this Indic perspective stresses the ecological advantages of formal education and the polluting effects of more traditional rural practices, notably soil erosion and deforestation.

The green movement is primarily material in its focus. It is concerned above all with advancing human interests (collectively and individually) and in the process re-embedding humankind within a natural framework. Its spiritual sensibility is an adjunct, subordinate to political agendas that seek changes in the here-and-now, and whose future goals assume concrete forms. These are the very goals from which Jain Dharma advocates eventual escape. Ecological awareness might exert a powerful influence over the way Jains (individually and collectively) think and organise their lives, but it is ultimately subordinate to the 'higher' consciousness of *Moksha*. Crucially, Jain consciousness is directed towards the individual, green consciousness towards human society and 'the planet'. The point of convergence is not at the immediately obvious areas of environmental concern, where environmental activists campaign or Green politicians seek office. Instead, it is found within the intricate system of Jain logic and its multi-layered view of the individual and society. Here, there is the possibility for those schooled in the western intellectual tradition to think about the environment and politics in new ways. Equally, the doctrine of 'many-sidedness' or Anekantavada can lay the foundations for an environmental politics that is not derived from or answerable to western needs and values.

Bibliography

Benton, T. (1993) *Natural Relations: Ecoology, Animal Rights and Social Justice*, London: Verso.
Benton, T., ed. (1996) *The Greening of Marxism*, New York: Guilford Press.

Capra, F. (1997) *The Web of Life: A New Synthesis of Mind and Matter*, London: HarperCollins.

Chakrawertti, S. (2004) 'Literacy Rate: Jains Take the Honours', *The Times of India*, Mumbai (7 September 2004 edn).

Dobson, A. (2007) *Green Political Thought*, 4th edn, London: Routledge.

Dundas, P. (2002) *The Jains*, 2nd edn, London: Routledge.

Dundas, P. (2004) 'Beyond Anekāntavada: A Jain Approach to Religious Tolerance', in Sethia, T., ed. *Ahimsā, Anekānta and Jainism*: New Delhi: Motilal Banarsidass, pp. 123–137.

Fortey, R. (1998) *Life: An Unauthorised Biography*, 2nd edn, London: Flamingo.

Glasenapp, H. von (1991) [1942] *The Doctrine of Karman [sic] in Jain Philosophy*, Varanasi: P.V. (Parshvanarth Vidyashram) Research Institute.

Hofmeester, K. (2013) 'Shifting Trajectories of Diamond Processing: From India to Europe and Back, From the Fifteenth Century to the Twentieth', *Journal of Global History*, vol. 8, no. 1, pp. 25–49.

Jaini, P.S. (2001) *The Jaina Path of Purification*, 4th edn, New Delhi: Motilal Banarsidass.

Kothari, J. (2004) 'A Diamond is Forever', *Jain Spirit*, Issue 20 (September – November 2004), pp. 48–50.

Lum, K. (2014, 16 October) 'The Rise and Rise of Belgium's Diamond Dynasties'. Available at: www.theconversation.com. Accessed 26 July 2019.

Mardia, K.V. (2007) *The Scientific Foundations of Jainism*, 4th edn, New Delhi: Motilal Banarsidass.

Mardia, K.V. and Rankin, A. (2013) *Living Jainism: An Ethical Science*, Winchester: Mantra Books.

Meadows, D. and Meadows, D. (1972) *The Limits to Growth*, New York: Signet.

Meadows, D. and Randers, J. (2004) *Limits to Growth: The 30-Year Update*, Hartford, VT: Chelsea Green Publishing Co.

Meena Peters, M. (2014) 'The Remarkable History of the Diamond Trade', *Passage* (March – April 2014 edn), pp. 12–13.

O'Riordan, T. and Cameron, I. (1994) *Interpreting the Precautionary Principle*, London: Earthscan.

Rankin, A. (2018) *Jainism and Environmental Philosophy: Karma and the Web of Life*, London: Routledge.

Shah, A. and Rankin, A. (2017) *Jainism and Ethical Finance: A Timeless Business Model*, London: Routledge.

Tobias, M. (1991) *Life Force: The World of Jainism*, Fremont, CA: Asian Humanities Press.

Zimmer, H. (1969) *Philosophies of India*, new edn, Princeton, NJ: Princeton University Press.

3 The Jain theory of pluralism
Transcending the politics of protest?

The hidden gem

Jain Dharma is based above all on the private expression of faith and the personal cultivation of wisdom. And, yet, as we have seen, its ethical doctrines contain a profound social dimension. They affect the way in which devotees study, conduct business and view the world, sometimes governing the most intimate details of their lives. In particular, Dharmic teachings influence the individual's view of his or her obligations to the surrounding society. In this context, we should remind ourselves that 'society' is defined as inclusive of all species, along with the environment that connects all 'embodied' life forms. Jains are guided in their daily lives by the Three Jewels, or 'Triple Gem', that summarise the principles of their faith: *Samyak Darshana* (Right Vision or Viewpoint), *Samyak Gyana* or *Jnana* (Right Knowledge or Understanding) and *Smayak Charitra* (Right Action or Conduct). Beneath these jewels, known as the *Ratnatraya* or *Triratna*, lies a fourth gemstone, hidden from view: the doctrine of many-sidedness or Anekantavada, which gives character and meaning to the principles and practices of the Jains and also connects together the concepts embodied in the Five Vows.

The reasoning process of Anekantavada is likened to the experience of light viewed through the facets of a cut diamond. Each facet represents an aspect of knowledge and the overall clarity reflects the elusive absolute truth. In homage to this analogy, there is even a company in Jaipur called Anekant Diamond Products. Jain Dharma is private and inner directed, but at the same time has strong public or social implications. It is also simultaneously relativist and universal in its perspective. There is a universal or absolute truth, which is identical with the Dharma as 'cosmic order'. However, the spiritual journey is defined as an attempt to grasp the truth and only those at the most enlightened levels can attain more than flawed or incomplete knowledge or grasp more than a few facets of reality. At the material level, no idea or belief system is absolute. As a philosophical method, therefore, the purpose of the doctrine of many-sidedness is to build upon the foundations of an unfinished 'edifice' of knowledge. According to Umasvati's *Tattvartha Sutra* (v.5.31):

> There is no viewpoint that is perfect and there is no science that is complete.
> ...[All] philosophies are imperfect although they are the glorious blocks that build the grand edifice of philosophy. ... And as there can be reality that science does not encompass, so there can be problems that are not solved by philosophy that is an endless quest.
>
> (Tatia 1994: p. 139)

Founded on the sense that each living being (or 'soul') has its own distinctive voice and viewpoint, this doctrine of multiple viewpoints reminds us of what we have in common with our fellow humans and the non-human world. It expresses the apparent paradoxes of the Jain belief system: the creative tension between continuity and change; the goal of transcendence and the practice of engaging with the world, and the equal emphasis on individual responsibility and the need for co-operation. At the same time, many-sidedness offers a way to resolve these paradoxes. This aspect of Jain philosophy is the point where all others intersect. There is a sense in which many-sidedness defines Jain Dharma and marks it out from rival points of view. From this perspective, Anekantavada is a particular response to the unique experiences of a minority population of adherents of an ancient faith tradition. Finding themselves surrounded by followers of larger, more politically powerful ideologies, they become at once tolerant of other viewpoints and defensive of their own cherished beliefs and way of life. In proclaiming their refusal either to seek converts or impose their ideas on others, Jain communities can appear to integrate with the outer world while simultaneously retreating into themselves.

Many-sidedness is a convenient way for Jains to face at once outwards and inwards at to seek integration with the wider society in which they live and work, but also safeguard their separateness from it. Viewed from an alternative angle – for with this doctrine everything is about angles – Anekantavada is the feature of Jainism that has potentially the greatest universal appeal as a 'unique selling point' in the market of ideas, with particular relevance to twenty-first century concerns. How, for example, do we reconcile the trend towards global interconnectedness through economics, popular culture, technology and movements of population with a contrasting reassertion of particular interests, including nationalism, religious faith and the resurgence of indigenous cultures? How, in multi-cultural societies, do we establish shared values founded on mutual trust between communities with contrasting and sometimes conflicting beliefs? How do we reintegrate humanity with 'the rest' of nature and at the same time preserve (and extend as widely as possible) the advantages conferred by education, medicine and relative freedom from nature's caprices? Such questions mirror the paradoxes inherent in Jain philosophy itself. Many-sidedness provides a form of logic that is likely, if not to resolve them in full, at least to address them with a renewed clarity and relevance.

Anekantavada is the Sanskrit term for many-sidedness and is usually contracted to Anekant. It is mentioned less often by Jains than Ahimsa, the guiding principle of non-injury, or even Aparigraha, the process of casting off possessive impulses and applying accumulated wealth to socially useful purposes (see Chapter 4). Nonetheless, it exerts a powerful influence over both practices. As well as being a doctrine or philosophical system, Anekant is (like 'Jainness') a sensibility, an attitude of mind that accepts multiple possibilities, the multi-layered nature of reality and the unique '*naya*' or viewpoint possessed by each form of life. The philosophical Anekant is arrived at through intellectual speculation or scientific inquiry. The sensibility of Anekant, by contrast, is achieved through lived experience, not least through the conduct of business and bringing it into line with the Five Vows (see Chapter 4).

The many-sided worldview can be viewed as a form of meditation, a de-cluttering of the mind of prejudices and preconceptions likely to induce a one-sided view (*Ekant*) and the actions that stem from it. The aim of these meditation exercises is to achieve a sense of perspective. This can lead to the actions associated with Aparigraha, if possessive attachment is understood as a form of one-sidedness that refuses to acknowledge the temporary nature of material success. Anekant reinforces Ahimsa, for it reminds the meditator that all life has value and purpose and that the powers we enjoy as human beings are cancelled out by our insignificance within the universe. The doctrine provides an exercise in intellectual Ahimsa or a 'non-violence of the mind' (Rankin 2006: pp. 159–93). Reality, which is identified with cosmic order or Dharma, has multiple aspects with differing degrees of visibility and obscurity, manifesting or concealing themselves according to circumstance and context. Existence itself is composed of 'origination, cessation and persistence' (Tatia 1994: pp. 134–40). The process of reinvention is eternal, in that although individuals and objects perish, every aspect of the inhabited universe is recycled and emerges in a new form, including the cosmos itself when it enters its next cycle. The future is 'endless' and the past 'beginning-less', but nothing within that framework remains static. The 'non-omniscient person' cannot 'perceive the existent in its reality':

> At a single moment, he can be aware either of the persisting unity (*ekatva*) of the substance [or object, person, animal, plant, etc.] and the transient multiplicity of its modes.
>
> (Jaini 2001: pp. 90–1)

Anekant does not provide a key to instant omniscience, but it allows for a greater understanding of the different levels or 'modes' of reality and the balance of continuity and change that governs all aspects of existence. At the intuitive level, it creates the sense that as living beings we are 'all in this together'.

Intellectual meditation: Anekant as an attitude of mind

Many-sidedness is an attitude of mind or frame of reference that guides daily decision-making. In this respect, it is based on 'experience and realism' (Singh, in Shah 2000: p. 126). The origins of Anekant are also found in complex scholarly discourses that have little direct relevance to the lives of most Jain householders but exert a powerful background influence. The abstruse theory of karmic accumulation translates into the practice of Aparigraha and forms the basis for a culture of philanthropy. In much the same way, Anekant's subtle and varied exploration of the nature of reality has given rise to a culture of tolerance. This form of tolerance does not imply easy or apathetic compromise between competing ideas. Instead, it involves acceptance that alternative viewpoints deserve critical evaluation and that ideas should prevail solely because of their intellectual worth and practical utility. Anekant is an inclusive doctrine of 'both/and' in place of 'either/or'. At the same time, it is a philosophy of 'perhaps' and 'maybe' that operates through a process of perpetual questioning and qualification rather than asserting certainties or even seeking to prove the validity of 'facts'. What is invisible is as important as what can be grasped and perceived. *Anekantavadins* (proponents of many-sidedness) admit of an infinite variety of possibilities, but simultaneously emphasise the limits of human knowledge, and hence the amount that we do not and probably cannot know.

The conceptual framework of Anekant arose out of manifold attempts by ascetics and scholars to explain the nature of reality, whether visible or invisible, finite or eternal. The result of these speculations was 'a denial of absolute existence or absolute impermanence' (Tatia 1994: pp. 134–40). Existence is thus neither a continuous state of flux nor a state of rigid immutability. It is changing all the time and these changes are reflected in continuous cycles of progress and regression, expansion and contraction, decay and renewal. This process of change is balanced by a sense of continuity or underlying stability. A person or an object might change with age and surrounding conditions but will retain an enduring identity as that person or object. This example is only partially effective because the human 'person' eventually dies and the natural or crafted object can erode, decay or be destroyed. Continuity in this context is not 'eternal' as it is with the life principle that unites us and at the same time makes us distinct. In the case of the jiva, the animating principle of continuity moves from one embodiment to the next until the point of escape from karma when it continues to exist in an unadulterated form. In so doing, it affirms the connections that bind together everything in the natural world.

This view of reality corresponds to the cyclical conception of time, divided into progressive and regressive half-cycles. That which is real (*tattva*) has two aspects, the eternal and the non-eternal or transient. It is permanent 'with respect to its essential substance and impermanent with respect to [the] modes [of existence] through which it is ceaselessly passing' (Tatia 1994: pp. 134–40). One 'mode' (that is to say a phase or aspect of existence) can be grasped at the expense of others'. The 'grasped mode' is 'brought to light' while other attributes remain in the background (Tatia 1994: pp. 134–40). Here we can draw certain parallels with more familiar forms of dualism, such as the polarities of Yin (darkness, nurturing, contraction) and Yang (light, activity, expansion) in Daoist and other classical Chinese thought (Cooper D. 2012; Cooper J. 1981). In Jain cosmology, there are similar creative shifts between solidity and flux, but with an emphasis on multiplicity rather than duality. Several propositions can be simultaneously regarded as 'true', not because they represent the 'whole truth', but because they reflect varying aspects or different 'facets' of reality.

It would be easy to confuse this approach with a form of extreme scepticism that accepts all modes of thought as equally valid. In fact, *Ajnanavada* or scepticism is explicitly rejected by the *Sutrakrita* (or *Sutrakritanga*), the second *Anga* (limb) of the Jain canon, as one of the paths 'opposed to the Jina'. The other paths, also described in unequivocal language as 'wrong' or 'false', include *Niyativada* (fatalism), *Akriyavada* (non-action), *nityavada* (eternalism) and *charvaka* (annihilationism: the idea that there is 'nothing beyond the senses' and so only material beings and objects exist). All of these are versions of *ekantavada*, one-sidedness, and are 'thus inferior to the comprehensive (*anekanta*) Jaina view of reality' (Jaini 2001: p. 53). The *Stotras*, a series of poetic hymns to the Jinas, also criticise '*ekantavadins*' who 'hold absolutist doctrines' and so are opposed to the 'doctrine of manifold aspects' (Jaini 2001: pp. 85–6).

Here there is an apparent contradiction between the acceptance of 'manifold aspects' and the rejection of competing doctrines as 'false'. Such reasoning is especially alien from the perspective of those schooled in conventional western logic, which is primarily concerned with choosing between opposing versions of the truth and emphasises the importance of consistency. Alternatively, the western method of argument aims to combine elements of contradictory propositions into a synthesis, which is viewed as a more complete version of the truth. The doctrine of many-sidedness seems, when viewed from this perspective, to be incapable of viewing some ideas or philosophical perspectives as erroneous. Moreover, a many-sided interpretation of reality does not seem so much to be a synthesis as a kaleidoscopic or psychedelic vision. Yet to understand Anekant, it is necessary to abandon the preconceptions implied by a narrow search for consistency. For falsehood is held to arise not from a 'false' idea or proposition itself, but from the refusal to acknowledge other possibilities and in that refusal embracing a closed system of thought. A fatalistic position (*niyativada*), for example, fails to take account of the importance of individual or collective endeavour. At a social level, it tends towards promoting apathy and discouraging constructive reform. Open-ended scepticism or extreme relativism (*ajnanavada*) is held to undermine the sense that there are absolute truths such as the centrality of Ahimsa.

Non-action is the state enjoyed by liberated souls who are depicted in Jain iconography as seated at the top of the universe, having floated there as soon as they are released from *samsara*. For the 'non-omniscient' person or being, however, the failure to act (and think carefully before, during and after each action) blocks the quest for knowledge and also prevents the accretion of positive karmas. It follows that noon-action is effectively the opposite of Careful Action. Similarly, the definition of 'eternalism' (*nityavada)* as a form of false consciousness would seem to contradict the idea of the universe as eternal, the product of energies that can neither be created nor destroyed. Yet eternalism in this context is defined as the belief that only the 'substance' or material aspect of a being or an object really exists, thus denying the 'modal aspect' which is insubstantial and can transfer from one incarnation to another. Its opposite, *anityavada* or 'non-eternalism', denies the literal existence of substance (*dravya*), accepting only the modal or transitory aspect of existence as 'real' (Jaini 2001: p. 92). This line of reasoning was associated (with some degree of accuracy) by early Jain scholars with Buddhism of the Theravada school, which today predominates in Sri Lanka, Thailand, Cambodia and Burma/Myanmar. The 'denial of substance', according to Jain logicians, 'makes it impossible ... to explain logically either bondage by karma (*samsara*) or the release from this bondage (*nirvana*)' (Jaini 2001: pp. 92–3).

Such arguments could be said to resemble the disputes that characterised early Christianity (Williams 2002). They arguably continue to play as important a role in the way Jain thinking today as in the age of Mahavira. The quest for knowledge is interpreted as a search for completeness, an effort to put together a universal jigsaw when many of the pieces are missing or hidden from view. However, the existence of missing pieces compels an element of doubt. This derives from an underlying awareness of the partiality and incompleteness of human knowledge (impeded by karma). From the premise of partial or qualified definition emerges the way of thinking referred to as *Syadvada*. The word *syat* in Sanskrit means 'might be', which is why *Syadvada* is at times referred to as 'maybe-ism'. For Jains, it is used to convey the idea of 'in some respect' (Jaini 2001: p. 94). *Syat* is further qualified by the word *eva*, literally meaning 'in fact'. In the context of *Syadvada*, the use of '*eva*' conveys the 'fact' as experienced by the speaker or writer, as opposed to alternative *naya* or interpretations:

> Thus, the statement 'the soul is eternal', when read with syat and eva would mean: 'In some respect – namely, that of substance and not of modes – the soul is in fact eternal'. By qualifying the statement in this manner, the Jaina (*sic*) not only makes a meaningful assertion but leaves room for other possible statements (for example, "it is not eternal") that can be made about the soul.
> (Jaini 2001: p. 94)

The soul (jiva or life monad) is eternal in that it never ceases to exist and so its 'substance' remains. However, it changes its 'mode' with the character of karmic influence it experiences and the cycles of incarnation through which it passes. It is, like everything in the Jain universe, at once permanent and impermanent, static and adaptive, depending on context. *Syadvada* is a way of navigating between competing or sometimes conflicting *naya*, not by merely conceding an argument but by admitting that multiple perspectives can coexist and that any proposition can be viewed from two or more angles:

> The spirit of this approach guards [the practising Jain] at all times from extreme viewpoints. ... Jainas (*sic*) are encouraged to read extensively in the treatises of other schools [of thought] ... It also seems likely that the failure of any doctrinal heresy to appear during nearly 3,000 years of Jaina tradition can be largely attributed to this highly developed critical analysis and partial accommodation.
>
> (Jaini 2001: p. 96)

There is a cyclical relationship between *Syadvada* and Anekant. The constant application of '*syat*' and '*eva*' induces a mood of perpetual questioning, reflection and reassessment, along with openness to new ideas or alternative viewpoints. The belief in 'many paths' towards the same ultimate goal of understanding reality fosters a spirit of inquiry and self-examination. It also, crucially, encourages an attitude of intellectual modesty. This helps us to understand why, as Padmanabh Jaini suggests (*op. cit*, p. 96), there has been no 'heresy' or radical schism in Jainism since its emergence as a distinctive spiritual tradition. There are two significant 'schools' of Jain practice, the *Svetambar* ('white-clad') and the *Digambar* ('sky-clad'): in the case of the former, the ascetics (male and female) wear white whereas the latter's most senior male ascetics practise nudity. The Digambar Jains are, arguably, somewhat more rigorous and uncompromising in their interpretation of the Jain way of life. Be that as it may, the two schools accept each other's differences without regarding themselves as rivals and often share facilities, especially in the Diaspora. Within both *Digambar* and *Svetambar* traditions, there are differing modes of thought following different practices or customs (for example those who use icons as meditational tools and those who reject their use), but again there is no sectarianism, merely a pragmatic agreement to differ.

The spirit of questioning is linked to the well-known sense of modesty found within Jain communities. Deeply held beliefs are rarely asserted or even widely discussed in public. This spirit is reflected in the absence of a missionary tradition and an attitude of equanimity towards uncongenial cultural practices or attitudes. Furthermore, the mental processes of Anekant and *Syadvada*, and the cautious disposition associated with *syat/eva* implicitly promote environmental awareness, although they are not directly associated with the more explicitly environmentalist areas of Jain philosophy. Careful thought becomes part of the practice of the Careful Action principle that guides all aspects of Jain behaviour. Treading lightly at the intellectual level serves as a reminder to 'live lightly on Earth', in the physical realm (Schwartz and Schwartz 1998), respecting the *naya* of animal and plant species and recognising human dependence on them. Anekant has indirect ecological implications through its influence on the way in which Jains relate to their surroundings, plan their lives and view society as more than a collection of individual human beings.

Anekant's relevance beyond the Jain communities where it evolved is open to debate, or more appropriately a balance of competing *naya*. A culturally distinctive phenomenon, it nonetheless issues an implicit challenge to all forms of exclusivity. The process of mental meditation through which a many-sided perspective is reached involves casting off certainties, including all notions of cultural superiority or 'ownership' of ideas. Anekant can therefore be two things at once. First, it is a central, defining feature of the Jain philosophy with a direct impact on the way in which decisions are made in daily life. Secondly, it is a method of intellectual and philosophical inquiry open to all who wish to make use of it. At the heart of the doctrine of multiple viewpoints, there is a radical critique of human supremacy over other species and the rest of nature by virtue of superior intelligence. This resonates with the concerns of political ecology and so should be of interest to those who wish to reassess the place of humanity within nature.

How green is Anekant?

In the United States, the Green Party of California expresses its key values in the form of questions. The principle of Ecological Wisdom, for instance, prompts supporters to ask, among other things: 'How can we operate human societies with the understanding that we are part of nature, not on top of it?'; 'How can we build a better relationship between cities and countryside?' and 'How can we guarantee the rights of non-human species?' Grassroots Democracy, another key value, invites questions such as 'How can we develop systems that allow and encourage us to control the decisions that affect our lives?' and 'How can we encourage and assist the "mediating institutions" – family, neighborhood organization, church group, voluntary association, ethnic club – to recover some of the functions now performed by the government?' We are encouraged to ask by the Postpatriarchal Values section (this key value encompasses both feminism and new perspectives on masculinity): 'How can we replace the cultural ethics of dominance and control with more cooperative ways of interacting?'; 'How can we encourage people to care about persons outside their own group?' and 'How can we learn to respect the contemplative, inner part of life as much as the outer activities?'. The other key values are Nonviolence, Decentralization, Community-Based Economics, Respect for Diversity, Personal and Social Responsibility, Global Responsibility and Future Focus (Radical Middle 2019).

Far from serving as a substitute for policy-making, this interrogative process serves as the preamble to a more detailed programme. In the context of American third-party politics, California's Greens have had a measure of success at the ballot box, electing mayors in a wide range of communities located for the most part in northern and rural areas of the state. The party's cultural influence has exceeded its polling figures and crosses the conventional right-left, conservative and liberal divides. In this sense, the California Greens are a useful case study in many-sided thinking applied to environmental politics. The reach into rural and predominantly conservative districts marks a difference with the Green parties of Europe. These have been primarily urban movements, whose adherents have experienced a sense of from the 'natural world' outside the city and are responding to the pressures of pollution and overcrowding on the quality of their lives. Many Californian Greens, by contrast, have been influenced by the counter-culture of the 1960s, from which in the subsequent decades evolved significant interest in alternative views of economics and spirituality. This gave rise to a 'back to the land' movement expressed through rural communes and distrust of centralised authority, whether political, corporate or technological (Rorabaugh 2015: pp. 205–27).

Asian philosophies have exerted a powerful and lasting influence over the counter-culture and its political manifestations, both directly and by osmosis. The civil rights movement, with its techniques of non-violent opposition to racial and social injustice, owes much to Mahatma Gandhi's *satyagraha* (truth struggle) against British colonial rule. The Mahatma, a Gujarati Hindu, was influenced in turn by a Jain merchant-philosopher, Shrimad Rajchandra (Shah and Rankin 2017: p. 95). Rajchandra encouraged him to consider the value of Ahimsa and use it to address social inequalities, including those arising out caste discrimination, since inequality and discrimination were in and of themselves forms of violence or *himsa*. Gandhi's economic model for India, *swadeshi*, was based on local production for local needs and a system of an economics that served the whole of the community rather than centres of affluence and political power

(Rankin 2010: pp. 134–5). Significantly, the village was viewed as the centre of the economic order, drawing upon and modernising traditional structures such as the extended family and the community 'bound together' (in Jain terms) by shared values. Technology was to be attuned to the environment and operate on a human scale. The spinning wheel became a symbol both of sustainable technology and economic self-sufficiency. The Sanskrit word *swadesh* means 'own country', with '*swadeshi*' denoting 'of one's own country'. This was part of a resurgence of national consciousness, but at the same time it pointed towards an ideal and an aesthetic associated with self-sufficiency, individual and communal responsibility, simplicity of clothing and lifestyle. Such concepts were later to be embraced with vigour by the counter-culture in North America, Europe and Australia (Roszak 1995) and underlie the key values of the California Greens and the questions stemming from them.

The long-term significance of the California Greens' mindset and political strategy has yet to be seen. It offers a useful example of how many-sided logic might be applied to a political discourse largely characterised by adversarial positions and doctrinaire certainty. By using the technique of questioning, it is possible to blur the conventional (and unsatisfactorily simplistic) divisions of 'left' and 'right' and find unexpected common ground with those who would usually view political and social issues from different perspectives. In the case of the Greens, it helps to explain why, their background in left-wing campaigns and the counter-culture notwithstanding, they have been able to adapt to cultural conditions and thus garner support in some of the state's most socially conservative rural regions. Another explanation is that numerous supporters of the counter-culture settled in these areas as a reaction against urban life. Their story of gradual and sometimes difficult assimilation also illustrates a 'many-sided' convergence of radical and conservative worldviews, as the incomers and established communities lived together, worked together and influenced each other often in subtle and unplanned ways (Rorabaugh 2015: pp. 205–27).

The political landscape in which the Ten Key Values arose is very different from that in which Anekant and *Syadvada* evolved. There is no conscious connection between them except for the indirect 'Asian' influences, although those owe more to Zen Buddhism, both in its 'pure' Japanese form and in varieties adapted to a blend of western consumer culture and green consciousness (Snyder 1999; Suzuki 2010 [1959]). That said, there is a very striking resemblance between the values enunciated by the California Greens and the principles of

Anekant. Both recognise that an attitude of questioning is a more practical response increasing complexity in economics, the environment and technology than adopting absolutist positions. Jains in India, and later in disparate immigrant communities, formulated the doctrine of many-sidedness as much as a survival mechanism as an exercise in speculative thought. Their participation in the economic life of India and the many Diaspora nations has been reliant upon a process of perpetual and careful negotiation, balancing the need – and desire – to integrate and participate with a jealous guarding of their distinctive identity. In a world of increasing cultural convergence, and at the same time heightened awareness of cultural distinctions, the methodology of Anekant is surely worth considering outside its original Jain context.

Nowhere, perhaps, is the logic associated with 'how can we' more relevant than in humanity's relationship with the environment. Anekant is not an environmental philosophy in the restrictive sense, but part of a view of society that encompasses all living beings, not just humans, and thus reintegrates humanity with nature. Furthermore, the idea that all forms of life have their own viewpoint that is worthy of respect and has its own 'intrinsic value', irrespective of its relationship with humankind, accords with the Deep Ecology platform enunciated by Norwegian Arne Naess and George Sessions, a Norwegian and an American philosopher respectively. Both men have been moulded in their thinking by the landscapes and historical conditions of their respective countries. The main theme of their platform is respect for the value of individual life, whether human, other animal or plant, combined with acceptance of the environments inhabited by humans as autonomous entities, with the same rights to self-determination and freedom from unwarranted interference as human individuals, communities and cultures. For example, the Deep Ecology platform includes the proposition that 'Richness and diversity of life forms ... are values in themselves' which are 'independent of the usefulness of the nonhuman world for human purposes' (Naess and Sessions 1984).

Despite its positive engagement with Asian and Native American philosophies, Deep Ecology remains profoundly rooted in the western cultural canon. It is at once a development of and a reaction against the principles of the European Enlightenment and the assumptions of industrial society. Its aim is to re-engage with nature, 'rather than adhering to an increasingly higher standard of living' (Naess and Sessions 1984). For humanity and the rest of nature, over-development has iniquitous social and environmental effects, including psychological and spiritual malaise. Reconnection with nature is in itself a form of 'liberation' for human beings, whether as individuals or collectively (Devall 1990; Devall and Sessions 1985). Connections are drawn between unequal or exploitative human relationships, in particular the economic and psychological oppression of women (Griffin 1984) and the exploitation of nature by humanity as a whole, including the oppression of non-human species. A revised and non-exploitative relationship between humankind and the rest of nature is viewed as the key to unlocking human potential (Naess *et al.* 2010), even more so than the transformative economic and social policies emphasised by eco-socialists, who make connections between exploitative class relationships, capitalist production (with its mechanistic emphasis on growth) and the exploitation of the environment, which becomes a mere resource (Benton 1996; Ryle 1988). Deep Ecologists of the Naess-Sessions school is open to aspects of the eco-socialist viewpoint, but points to the poor records of socialist economies in the ecological sphere (Naess *et al.* 2010). Supporters of the green left who are influenced by Deep Ecology are quick to point out that socialist programmes imposed by authoritarian means very quickly adopt some of the worst features of capitalist production (Bahro 1982; Pepper 1993).

All this is far removed from the Jain ideal of individual self-realisation through withdrawal from the constraints imposed by nature, although the starting point is tantalisingly similar, recognising the interconnectedness and mutual dependence of all forms of life. Like Anekant, Deep Ecologists (and most eco-socialists) also recognise the importance and 'intrinsic value' of each living being, including its distinctive viewpoint. Deep Ecology and associated schools of thought such as ecofeminism (e.g. Griffin 1984; Plumwood 1993) represent a branch of western environmental philosophy that aims radically to redress the balance between humanity and the natural world. However, *Anekantavadins* differ from Deep Ecologists in two important respects. First, they operate within the framework of Jain Dharma. It cannot be overemphasised that the faith-based aspect of Dharma is primarily concerned with escape from the natural world, along with all material attachments, which are manifestations of karma. Return to nature in an idealised form is not compatible with this aim. Jain ascetics, who come closest to the ideal of material renunciation, are not seeking to be 'at one' with the natural world but to escape as far as possible its demands, because they do not draw a (western-style) distinction between nature and the material demands of urban living.

Secondly, those lay Jains who actively attempt to incorporate Anekant into their daily lives do not idealise the natural world. Unlike Deep Ecologists, Jains who are concerned with environmental issues are all too aware of the imperfections of vicissitudes of 'pristine nature' or 'nature in the raw'. Those who are actively involved in philanthropic work are as interested in alleviating problems with a 'natural' cause as those with a 'human' origin. In many cases, notably droughts, flooding and climatic instability, environmental crises arise when human carelessness pushes to breaking point already delicate ecosystems. There is also a profound environmental impact arising from the poverty, urban and rural, which both Jains and western ecologists recognise (Gerrard 1995; Martinez-Alier 2002), although the former do not romanticise poverty and subsistence living, except as an ascetic ideal. Moreover, the Indic origins and bias of Jain thought balances western environmentalists' concerns about over-development about the equitable distribution of resources.

The work of Veerayatan, described in more detail in Chapter 4, includes both environmental and technological education in its poverty reduction programmes for rural Bihar. This many-sided approach accepts that nature, in particular natural disasters, can be a major contributor to poverty. Education and the sensitive application of technology improve understanding and management of the environment, making natural disasters less likely. The accoutrements of modern living, such as running water and homes that withstand the elements, improve the quality of human life. They can also enhance, rather than undermine, the relationship between humanity and the natural world, replacing a sense of continuous struggle 'against nature' with a desire to conserve, work with and identify with it as much as possible. The emergence of the green movement in the most 'developed' economies, perhaps, is evidence that prosperity can at least at times induce a more constructive engagement with nature. Furthermore, improvements to the quality of life and the raising of expectations beyond subsistence level are part of a larger process of opening educational and career opportunities and looking beyond the short-term need to survive 'hand to mouth' from day to day. In a literal sense, they give people time to think. They are thus more open to looking beyond immediate interests and considering a wider range of possibilities.

There is, in this sense, a contrast within Jain practice between the *Mahavrata* objective of non-attachment pursued by the highest levels of ascetic and the *Anuvrata* objective of reform, pursued by 'civilian' men and women and some (especially female) ascetic orders. The first is highly personal and its premise is escape from the environment. The second is social and involves positive engagement with the environment and a commitment to improving the human condition. It is recognised that poverty and underdevelopment create and intensify attachments rather than eroding or abolishing them. These attachments can translate into fanaticism and violence, including violence against nature in the interests of immediate survival. The amelioration of such conditions is, despite appearances to the contrary, fully compatible with the vow of Aparigraha. 'Non-possessiveness' involves the sharing of and recycling of resources, as opposed to static concentrations of wealth and power.

It follows that, from the perspective of spiritual liberation, the renunciation of the world is not the sole objective, to be pursued in all situations and circumstances. Similarly, 'either/or' divisions are absent from the predominant Jain attitudes towards business and its relationship with the environment. The business ethos of Jain Dharma is founded on a rigorous ideal of social and personal responsibility and self-imposed restraint. It is not necessary to make a 'choice' between commercial and conservationist activities, but the way in which both are conducted is of critical importance, as are the underlying intentions. The multi-faceted or 'many-sided' perspective offers the possibility of an environmental politics that is neither economically regressive nor culturally Eurocentric and is defined by a pluralist respect for human and non-human viewpoints alike.

Bibliography

Bahro, R. (1982) *Socialism and Survival*, London: Heretic Books.
Benton, T., ed. (1996) *The Greening of Marxism*, New York: Guilford Press.
Cooper, D. (2012) *Convergence with Nature: A Daoist Perspective*, Totnes: Green Books.
Cooper, J.C. (1981) *Yin and Yang: The Taoist Harmony of Opposites*, Wellingborough: The Aquarian Press.
Devall, B. (1990) *Simple in Means, Rich in Ends: Practicing Deep Ecology*, London: Green Print.
Devall, B. and Sessions, G. (1985) *Deep Ecology: Living as If Nature Mattered*, Layton, UT: Gibbs Smith.
Gerrard, B. (1995) *Whose Backyard, Whose Risk: Fear and Fairness in Toxic and Nuclear Waste Siting*, Cambridge, MA/London: MIT Press.
Griffin, S. (1984) *Woman and Nature: The Roaring Inside Her*, 2nd edn, Berkeley, CA: Counterpoint.
Jaini, P.S. (2001) *The Jaina Path of Purification*, 4th edn, New Delhi: Motilal Banarsidass.
Martinez-Alier, J. (2002) *The Environmentalism of the Poor: A Study of Ecological Conflicts and Valuation*, Cheltenham/Northampton, MA: Edward Elgar.
Naess, A., with Drengson, A. and Devall, B., eds (2010) *The Ecology of Wisdom: Writings by Arne Naess*, Berkeley, CA: Counterpoint.

Naess, A. and Sessions, G. (1984) 'The Deep Ecology Platform', Foundation for Deep Ecology. Available at: www.deepecology.org/platform.htm. Accessed 29 July 2019.

Pepper, D. (1993) *Eco-socialism: From Deep Ecology to Social Justice*, London: Routledge.

Plumwood, V. (1993) *Feminism and the Mastery of Nature*, London: Routledge.

Radical Middle (2019) 'Ten Key Values' [of the California Greens and some other US Green Party Chapters]. Available at www.radicalmiddle.com/ten-key-values.htm. Accessed 29 July 2019.

Rankin, A. (2006) *The Jain Path: Ancient Wisdom for the West*, Winchester: O Books.

Rankin, A. (2010) *Many-Sided Wisdom: A New Politics of the Spirit*, Winchester: O Books.

Rorabaugh, W.J. (2015) *American Hippies*, New York: Cambridge University Press.

Roszak, T. (1995) [1969] *The Making of a Counter-Culture*, Berkeley: University of California Press.

Ryle, M. (1988) *Ecology and Socialism*, London: Radius.

Schwartz, W. and Schwartz, D. (1998) *Living Lightly: Travels in Post-Consumer Society*, Charlbury: Jon Carpenter.

Singh, R. (2000) 'Relevance of Anekānta in Modern Times', in Shah, N.J, ed. *Jaina Theory of Multiple Facets of Reality and Truth*, New Delhi: Motilal Banarsidass, pp. 127–135.

Shah, A. and Rankin, A. (2017) *Jainism and Ethical Finance: A Timeless Business Model*, London: Routledge.

Snyder, G. (1999) *The Gary Snyder Reader*, Washington, DC: Counterpoint.

Suzuki, D.T. (2010) [1959] *Zen and Japanese Culture*, Princeton, NJ: Princeton University Press.

Tatia, N. (1994) *That Which Is: Tattvārtha Sūtra (Umāsvāti)*, San Francisco, CA: HarperCollins.

Williams, R. (2002) *Arius: Heresy and Tradition*, Grand Rapids, MI/Cambridge: SCM Press.

4 Jainism and environmental politics
A radical synthesis?

Loosening the karmic chains

One of the distinguishing and, for outsiders, often disturbing features of the Jain belief system is that it is radical in the literal sense. Problems, human or environmental, are examined from the roots upwards. The starting point, as we have explored above, is the individual entity, or jiva. The jiva's spiritual journey takes it through a range of experiences, positive and negative, towards a state of liberation (*Moksha*) or full enlightenment. It begins this process as a unit of pure consciousness and ends it in the same form but with a full understanding of the way the universe works. In between, the jiva experiences many 'embodiments' in human, animal and plant or sometimes, when viewed from a rational human perspective, supernatural forms. It thus possesses an underlying individual essence, its spiritual core, but assumes varied identities spanning the spectrum of life in the 'inhabited universe'. In terms of the quest for knowledge, which is the purpose of the jiva's existence, the transient assumed identities are as important as the original essence. Continuity alternates with change, within a worldview that is at once profoundly individualistic and strongly collectivist. The individual is responsible for his or her own choices and the impact they have on others, with the level of responsibility increasing in accordance with that individual's intelligence, education and power. At the same time, there is a dual concept of individuality: it is simultaneously static and fluid, self-contained and transferable between 'lifetimes' or incarnations. All life is interconnected (*Parasparopagraho Jivanam*) and yet each unit of life is ultimately in charge of its own destiny. Individual autonomy is supreme, but the aim of each individual is to adhere as closely as possible to the Dharma. Dharma, for Jains, is experienced as both the natural order of the universe and an ethical system by which we should live. Alignment with

Dharma involves following the Vratas, the most important of which is Ahimsa, which enjoins us to avoid injury to all forms of life. The radical element in the Jain worldview is expressed in two ways. First, the series of paradoxes or apparently contradictory propositions listed above would appear, from a western standpoint especially, to negate or at least avoid choice. Yet for Jains, it represents a deeper understanding of reality. Many of the 'choices' between individual interests and common good, rights and responsibilities, animal and human welfare are deemed illusory or 'one-sided' (see Chapter 3). A greater understanding of truth arises when the layers of dogmatic certainty are removed to reveal a plural, almost kaleidoscopic vision of reality, the essence of which is expressed through constantly shifting patterns. Ideas, including political ideologies, are seen as but fragments of the truth. To mistake them for 'the whole' truth is to exclude large areas of knowledge. Furthermore, it means engaging in a form of intellectual attachment, which for Jains is no different from *Parigraha*, or material possessiveness. Ideas cannot be 'owned'. Nor should they be adhered to with a rigid and exclusive fervour: this generates passionate anger (*kasaya*), fanaticism and 'self-deluding' forms of karma. The pluralism of this perspective contrasts with the prevailing adversarial mode of politics, including the self-righteous certainty often attributed to environmental campaigns.

Secondly, the Jain perspective is radical in its perception of matter and the different layers of reality. Matter that is without a spiritual component is defined as *ajiva*, devoid of 'real' life essence. Much of the material in the universe that at a mundane level of consciousness we associate with life is regarded as *ajiva*, including the environment in its material manifestations. Material that is *ajiva* is ultimately without meaning or value and has only temporary relevance. Karma plays a critical and perhaps the most important role in the Jain understanding of reality. As a process, it binds the individual jiva into the realm of material objects and the concerns that accompany them. Karmic particles themselves are a form of *ajiva* because they consist of non-living matter in an infinitesimal form. They vary in quality and density and are attracted to the jiva by all forms of action, from the initial vibratory movement (when it comes into existence spontaneously) to the lived experience of all its physical incarnations. The fully enlightened being – the jiva after *Moksha* – has become motionless: it is both free from karma and unable to attract karma. Karmic fragments of matter (*karmas*) are sometimes viewed through the prism of modern science as subatomic particles or 'karmons' (Mardia 2007: p. 10).

Jains, unlike adherents of many Hindu traditions, do not view the jiva as a tiny, subtle particle but as a more complex being:

> [The jiva] pervade[es] the whole organism; the body constitutes, as it were, its garb; the life-monad is the body's animating principle [while] the subtle substance of this life-monad is mingled with particle of karma, like water with milk, or like fire with iron in a red-hot, glowing iron ball.
>
> (Zimmer 1969: p. 229)

The awakening of consciousness and the pursuit of knowledge involve navigating the different levels of reality to determine what is of lasting value. Thought processes and actions are all in different ways karmic, however positive the motivation and outcome. Yet it is only through activities, intellectual and physical, that it is possible to see beyond the karmic cycle. Umasvati's *Tattvartha Sutra*, which distils and clarifies much of Jain thought for the lay practitioner, lists nine aspects of reality or 'things' (*tattva*), defining what is real in contrast to what is ultimately false or transitory. The *tattva* (or '*tattvas*') are sometimes referred to as the 'Nine Reals' (Tatia 1994: pp. 147–63). Some of these aspects of reality have been mentioned in previous chapters, but the list gives them a larger cosmological context. I have taken the designations of the 'Reals' from Umasvati, but adapted the commentary to reflect where possible the relationship with ecological concepts:

Jiva: unit of life, soul, that which is sentient and has the capacity for spiritual growth

Ajiva: inert matter, material as opposed to living 'things', that which lacks sentience or potential for enlightenment

Asrava: the influx of karma (karmic particles), obscuring the jiva's awareness of its true self (and blocking the individual's spiritual development

Papa: negative or destructive karmic influence (including negative thoughts and actions, violent fanaticism, false or misleading material attachments)

Punya: positive karma (including creative or altruistic actions including environmental conservation and animal welfare, benevolent or loving attachments or thoughts); *punya* activities are still karmic because all actions attract karma, but they point the way to self-realisation and release from karmic influence

Bandha: 'karmic bondage', or the experience of being encased in (particles of) karma and exclusively or overly involved with material concerns, leading to further karmic influx (Asrava)
Samvara: stoppage or 'closing off' of karmic influx; awakening consciousness
Nirjara: breakage, shedding, falling away of karmas/karmons and their influences; further development of consciousness
Moksha: liberation, release from *samsara* (the karmic cycle of birth, death and rebirth), omniscience or full understanding of reality, 'self-conquest' or realisation of the inner self (the jiva).

Interestingly, inert or 'soulless' matter, including karma itself, counts as 'real' although it obscures and complicates our understanding of the 'true' reality beneath the surface. Here it is unhelpful to think in binary, western-style terms of real and unreal, but to envisage different degrees or levels of reality. Understanding the workings of the universe and living in accordance with Dharma means learning to discriminate between what is important and necessary in the short or medium term and what is of lasting or ultimate significance. This process of discrimination corresponds to the difference between the essence of the self, the jiva, and the temporary 'selves': the many modes of the jiva on its journey towards full consciousness. Money and possessions, including property, are therefore classified as *ajiva* but are still 'existent', or real at a secondary level. It follows that material concerns, as varied in form as commercial activities and the protection of the environment, need not be ignored nor treated as irrelevant distractions. With the exception of ascetics, who have chosen to abstain from the ordinary rules of living to pursue lives of meditation and devotion, Jains are encouraged to live in and engage with their surroundings while keeping in mind their ultimate goal of transcendence.

Where material possessions become dominant or are held to be ends in themselves and the governing principle of life, they are defined as part of a 'deluded' view of the world. The term *Mohaniya* is applied to the karma leading to self-delusion. Karmic influence, which blocks understanding of the self and the world, attracts further karmas and so the jiva can easily be sucked into a cycle of ignorance (*avidya*). Only conscious choice can break that cycle. In practice, this means awareness of social obligations, defined by the Five Vows (*Vrata*) and encompassing obligations to the environment as part of the definition of social. Karma, especially 'heavy' or 'negative' karma, tends to replicate itself, but it also contains the seeds of its own destruction, by inspiring conscious resistance to its influence. Humans are (according to

Jain definition) creatures with the highest number of 'senses' (Mardia 2007: pp. 22–5) and the strongest capacities for conscious choice. This gives us, as human beings, an extra layer of spiritual obligations, which extend to the social and environmental arenas, in particular the obligation to refrain from harm to others. At the same time, the human capacity for delusion, and its destructive consequences, is significantly increased.

Many Jains refer to the deluded state as *Maya*, as form of deceit and delusion that stands in the way of *Darshana*: 'Right Vision' or viewpoint, in terms of devotional religious practice, 'correct view of reality' or 'true spiritual insight' in philosophical terms (Jaini 2001: pp. 151, 351). *Samyak Darshana* is essentially a state of mind, a disposition rather than a doctrinaire or self-righteous stance. It acknowledges multiple possibilities and perceptions of reality and makes the attempt (albeit never fully completed before Moksha) to fuse them into a unified whole. The quality of open-minded acceptance of a larger truth, or 'holistic worldview', is referred to as *astikya*, which is sometimes rendered in English as 'faith' or 'belief'. As only partially accurate translations, these terms can be especially confusing to the western student of Jainism, because they can imply uncritical acceptance or resignation. *Astikya* is better defined as an implicit understanding of the nature of reality, at once intuitive and rational. That understanding, in turn, is only partial. The disposition of *Samyak Darshana* requires intellectual humility: the pursuit of knowledge combined with an understanding that our capacities are limited. From such understanding it follows that 'one-sided' ideas or partial versions of the truth cannot be imposed on others or confused with absolute truth. Delusions of power are challenged at their roots. These include the right of humanity to exert harmful forms of control over the rest of nature as a result of (allegedly) higher capacity for reason than other species. *Samyak Darshana* also challenges the prevailing myths of human supremacy and invincibility. The true power conferred by human intelligence and spiritual knowledge is the power to avoid harmful actions.

Astikya, which leads to *Samyak Darshana*, is arrived at through devotional prayer, using the *Tirthankaras* as points of reference, or meditation to achieve a state of equanimity. It can also be achieved through study or lived experience, including charitable works and the ethical conduct of business or professional activities. It is an affirmation of reality, producing a state known as *Shraddha* or 'educated faith' (Jaini 2001: p. 151). Intuition, reinforced by reason and greater knowledge, gives rise to a state of mind that is at once rational and intuitive. Samyak Darshana is one of the Three Jewels or 'Triple Gem' of Jain Dharma (*Ratnatraya* or *Triratna*), along with *Samyak Gyana* [or *Jnana*] (Right Knowledge) and *Samyak Charitra* (Right Conduct). Knowledge itself is, as we have seen, often likened to a multi-faceted jewel (see Chapter 3). The *Triratna* balance an exclusively scientific understanding of the world with a mystical or devotional approach to nature and the universe. As inner states of consciousness, they are achieved through the sentiment of *Jiva Daya* (sympathy or identification with all life) and the sensibility of 'Jainness' (Mardia and Rankin 2013: p. 26) as much as by book learning. They impact, often subtly, on all areas of life and on the attitude that informs everyday activities. Knowledge, whether of an intellectual or technical kind, is also essential for dispelling delusion; this is one of the reasons for the traditional Jain emphasis on lifelong education for both men and women. The delusions leading to *himsa* are closely related to a state known as *Mithyatva* (or *Mithyadarshana*), false consciousness or false belief, a precursor to passions (*kasaya*) and one-sided viewpoints (*Ekantika*), including delusions of dominance or absolute knowledge. The regressive journey induced by *Mithyatva* is best described as a spiralling downwards into karmic influence and away from enlightenment. In this non-linear form of spiritual development, there is no guarantee of progress towards enlightenment. Jain Dharma is 'not a theory of necessary evolution; the Jaina (*sic*) also accepts the possibility of retrogression' (Jaini 2001: p. 111). An upward or downward spiral zigzagging through different evolutionary stages is accepted as the norm for the jiva's journey, and so evolution is not regarded as a straight line of inevitable progress.

In Jain terms, the belief that human beings are separate from their environments is a prime example of deluded karma or *Mithyatva*, as is the attendant idea of superiority, by which the environment is turned into a disposable 'resource' for human use.

Such notions are also extensions of *Parigraha*, the sense that we 'own' our natural surroundings and that their only value is that which can be assessed in monetary terms or in their susceptibility to human exploitation. The vow of Aparigraha, by contrast, offers a path towards *Nirjara*: the falling away of karmic particles mirrors the disposal of wasteful or unnecessary possessions, which are then put to constructive use (a form of *punya*). When the process of shedding karmas begins, the particles are said to fall from the subtle body of the jiva 'like ripe fruit' (Jaini 2001: p. 113). *Punya* is associated with enlightenment and is depicted as 'light' or 'white' karma in contrast to *papa*, which is portrayed as 'dark' or 'black' (for further discussion of *Leshya*, or karmic colours, see Chapter 2). *Punya* is also viewed as a physically lighter karma, yielding more easily, whereas *papa* weighs the subtle body down. Enlightenment and lightness of weight are both associated with reducing consumption and adopting a more detached stance towards the rest of the natural world. Nature is thus seen less as something to be 'possessed' and 'used', but more as a complex array of life forms and modes of existence, linked invisibly by the shared possession of jiva and a common involvement in the karmic cycle of *samsara*. More importantly, nature is no longer viewed as 'outside' or 'other' but as intimately connected to 'us' or part of our selves. Ironically, it is karma that produces this network of connections, from which the 'environmentalist' Jain motto *Parasparopagraho Jivanam* has arisen. Separation from karma means a state of full individual autonomy that few human beings would find attractive, and which is uninvolved with the processes of the natural world. Recognising shared possession of jiva as a form of common heritage connecting all beings in the inhabited universe also means recognising that we are all 'bound together' on the same journey through the evolutionary cycle. Attendant to this is an understanding that the viewpoints (*naya*) of others – especially other species – are valid and needs to be taken into account and wherever possible accommodated.

Paradoxically, the first steps towards loosening the karmic 'chains' and achieving equanimity are taken through social engagement. This is defined in wider terms than are habitually used within the prevailing western discourse. There is a powerful strain in political ecology, for instance, that is sceptical about commercial activity and regards the profit motive as inherently deleterious. In place of private ownership and market economics, alternative economic models, including co-operatives, are viewed as inherently more attuned to nature's rhythms (Bahro 1982; Benton 1993, 1996). Within green politics, there are some strong affinities with democratic socialism, tempered by an awareness that socialist economics have, for the most part, been as concerned with continuous economic expansion and committed to human supremacy over nature as their capitalist counterparts.

'Requiring minimum violence'

In the West since the mid-twentieth century, greens have at times looked towards the decentralist socialist and anarchist traditions of, for example, William Morris and Peter Kropotkin, the former inspired by an idealised version of the craft guild systems of Medieval Europe, the latter by the ideal of the self-governing peasant commune

(Kropotkin 2014 [1902]; Morris 1993 [1890]). Kropotkin's view of mutual aid, rather than the competitive drive, as the true motor of evolution has much in common with the ethical doctrines of lay Jains, including entrepreneurs who effectively recycle their gains through philanthropic acts. This outlook contrasts markedly with the radical individualism and asceticism associated with the spiritual doctrine of 'self-conquest'. Ideologies such as Gandhi's *swadeshi* movement in India are identified by mainly western greens with a decentralist model of 'eco-socialism'. *Swadeshi* is based on local production for local needs and a human scale use of technology attuned to the environment (Rankin 2010: pp. 134–5). Although it valued the role cooperative enterprises and prioritised communal welfare over profit, its basis was small-scale family businesses and its primary focus was local self-sufficiency. A strand in Jain thinking influenced *swadeshi*, which opposed dehumanising forms of industrialisation and urban growth, the prevailing model for 'development' in post-independence India, while emphasising non-violence, restraint and 'living lightly'.

Jain ethics, like *swadeshi* and the strong decentralist strain in political ecology, value co-operation and many Jain commercial enterprises are run on essentially co-operative lines. That said, a more flexible definition of social engagement is adopted than is characteristic of western greens. Commercial activity is often viewed as an innate force for good when it helps to lift individuals and communities out of poverty and promotes education, social (extending to animal) welfare and the protection of the environment. In many ways, the Jain position prefigures the idea of sustainable enterprise that recognises environmental responsibility as both ethically 'right' and economically efficient. Jain values go further than this, blurring or even erasing the distinctions between business and charitable activities, including nature conservation. This stance is reflected in the work of the Veerayatan movement, which we shall consider below, promoting modern commercial activities as a way of reducing poverty and protecting delicate natural environments so that they may continue to sustain life. Technology, in this context, is used to reconnect humanity and nature, providing greater security to the former and reducing harm to the latter.

Since all activity, whether physical, intellectual or commercial, is held to be to varying degrees karmic, it is in a sense beside the point to differentiate between different types of benign activity (*punya*), but more useful to focus on the intention behind the activity and its practical results for society (such as educational progress or the reduction of inequality) and the individual (such as the practice of 'non-possessiveness' and respect for multiple viewpoints). Short-term thinking, uncritical materialism, one-sided dogmatism in politics and the belief in human supremacy over nature are usually the origins of harmful activities (*papa*). Long-term planning that takes account of the 'binding together' of everything within the physical universe is more likely to produce positive results at both personal and social levels. Along with the interplay of continuity and change, Jain thought emphasises the creative interaction between the centrality of the individual, whose consciousness can evolve or regress, and the vastness of the universe, in a constant state of expansion and contraction. This gives the thoughtful Jain practitioner a sense of perspective, an understanding that all knowledge apart from the ultimate truth is relative and an inoculation against delusions of grandeur.

Jain Dharma is not a political ideology, but a spiritual philosophy, a cultural 'mentality' and a series of ethical codes for living in the world or (in the case of ascetics) achieving freedom from it. The reductionist labels and shifting frontiers of politics do not limit its scope and are often viewed as irrelevant. For this reason, practising Jains can appear to the observer as apolitical or 'small-c' conservative. This assumption, reinforced by the strong pro-business bias of Jain communities worldwide, is at best only a partial truth. The reality behind it is far subtler and more interesting. By applying the principles of the Five Vows to the problems of everyday life, many Jain men and women are in fact broadening the definition of politics. In the overlap between commercial activity, charitable giving and philanthropy, environmental conservation and animal welfare, there is an implicit political message. It is not articulated through slogans or grandiloquent gestures, but through practices embedded over generations. Yet by its very existence, what we might call the 'Jain model' presents us with an alternative way of looking at commercial activity, its purpose and long-term objectives. One of the main ways in which it does this is by breaking down the divisions between self-interested and altruistic endeavours. The latter is built into the commercial process rather than being viewed as supplementary, while the former is valued as a means to an end. The limits of both the self-interested and other-directed areas of human activities are freely acknowledged, because the eventual purpose of Jain ethics is individual salvation: freedom from the karmas of which all activities are part. The goal of transcendence remains a rarely articulated backdrop, and yet it provides a sense of perspective by which both utopian forms of altruism and uncritical obeisance to 'market forces' are avoided.

Jain philanthropy is offered, by convention, without ideological strings attached, or at least without the assumption that the beneficiaries will turn into obedient converts. In this sense, the Jain attitude differs markedly from those of political campaigners who seek allegiance and even a form of gratitude, or the predominantly western Non-Governmental Organisations (NGOs), which tend to be swayed by fluctuating political and cultural fashions within western societies. Many Jain charitable endeavours involve animal welfare, notably the *Panjrapoor* animal sanctuaries. It is made clear that little or no distinction should be drawn in terms of merit between human and non-human welfare. It is also because of the characteristic emphasis on freedom of conscience (a subset of Anekant, or acceptance of multiple viewpoints) and the exertion of indirect influence, by example rather than centrally imposed diktat. There is no need to 'become' a Jain or mimic rituals specific to Jain culture: it is enough to live out the basic principle of Ahimsa. So strong is this emphasis on freedom of thought that an important component of aid to human communities is conferring the power for individuals and communities to evaluate problems. They are encouraged to make their own decisions based on evidence and broad philosophical principles of non-injury to life. In this way, they are able to progress from mere subsistence to the ability to exercise choice. It is recognised that the culture of Jain communities is specific and distinctive. Upholding and perpetuating this distinctiveness is of the highest importance to practising Jains. When they reach out to the wider community, they are called upon to respect the culture and historical experience of others without imposing their own values. In general, Jain philanthropy makes explicit connections between concern for human welfare with the need to protect the environment. At the same time, it crosses the boundaries of species to maintain the commitment to interconnectedness. Educational and business opportunities are also both integral parts of the philanthropic process. The lines between them are blurred because education is viewed, for good reason, as a route out of poverty and dependency.

A significant example of this approach to welfare is the Veerayatan movement organised by Jain ascetic women in a region of Bihar, northern India, 'where Tirthankara Mahvira spent fourteen rainy seasons' (Shilapi, in Chapple 2002: p. 166). These female ascetics demonstrate that (as so often with Jainism), there is not always a clear division between the two types of vow. They practise the *Mahavratas* (Greater Vows) in their personal lives through dietary strictures, celibacy and renunciation of possessions, while directing a programme of social engagement that is in keeping with the *Anuvratas* (Lesser Vows) and the principle of *punya* (positive karma).

The emphasis of their grassroots environmental movement is on planting trees and educating rural communities in this impoverished area of one of India's poorest states. Villagers are shown and encouraged to think for themselves about ways to work better with rather than against the grain of nature, so that their environment is transformed into a positive force that is on their side, rather than an opponent to be worked against or subdued. The assumption it that a more beneficial relationship with the ecology of the region has beneficial effects on health, as well as contributing to the survival of long-established cultural patterns. Survival is achieved through evolution and adaptation to changed circumstances and not excessive traditionalism or stasis. Education in environmental management enables the ecology of the region to become gradually less precarious and more yielding to benign and carefully considered forms of human intervention. Meanwhile, local people are equipped with new skills to enable them to plan as far as possible for unexpected natural events such as storms, floods or droughts. The devastating effects of such episodes are made worse by unsympathetic human activities at local level in addition to global patterns of atmospheric pollution and climate change. Veerayatan's educational projects attempt to provide the local population with a countervailing force, in which non-injury becomes an active source of resilience and strength. Sadhvi Shilapi, a female ascetic, who has served for many years as Director of Education for Veerayatan in Bihar, expresses the social and educational aims of her movement in these terms:

> People are being given incentives to plant more trees. Drinking water, food, shelter, and employment facilities are being provided for the local population so that their dependence on the remaining natural resources for their livelihood is reduced. ... Villagers are taught [about] the protection of animals ... and the importance of protecting natural resources. ... The activism at Veerayatan is based on the universal principle of Tirthankara Mahavira that the sun, the air, the water and nature as a whole give of themselves silently and selflessly all the time. It would be selfish on our part if we take and do not at least return a portion of what we have taken by the time we leave the world. The mission works on this motto, given by Mahavira in ...the *Uttaradyayan Sutra* [Svetambar text]: 'Let friendship be our religion, not only in our thoughts but in our actions as well'.
>
> (Shilapi, in Chapple 2002: p. 166)

At a spiritual level, Sadhvi Shilapi also cites the *Adipurana*, a ninth-century CE text in honour of Rishava, the first Tirthankara who (in the spirit of many-sidedness) is also an aspect of Shiva, the revered Hindu deity:

> The text emphasizes that forests modulate the climate, check thunderstorms and floods, protect the neighbouring areas from cold winds, and enable the constant flow of rivers. ... The *Adipurana* says that forests are like saints, or *munis*, who, overcoming all obstacles, create a better welfare for all. ... As in the relationship between bride and bridegroom, it is the duty of all of us to protect and preserve the forest. To live a peaceful life and earn positive karma, the *Adipurana* suggests the planting of a tree. It is said that one who plants a tree remains steadfastly close to God.
> (Shilapi, in Chapple 2002: p. 162)

For non-Jains, the reference to 'God' is puzzling, since Jain Dharma is specifically non-theistic. There is no *Ishvara*, the personal god, Supreme Being or creative deity acknowledged by many schools of Hindu thought. Nor is there a concept of a god as First Cause. Instead there is only the interaction between the jiva and karma. The universe that progresses and regresses through long half-cycles and is constantly modified by *Kāla* (Time): this latter concept is also expressed in variant forms within Hindu and Buddhist cosmology. Jains, nonetheless, display tolerance towards other religious or cultural traditions and philosophical standpoints. In India, they participate in ceremonies and festivals in honour of local Hindu gods and goddesses, holding in high esteem many of the principal deities of the Hindu pantheon, including Shiva (Jain 1993). Saraswati (or Sarasvati), the Hindu goddess associated with learning, the quest for self-knowledge and the healing power of water is revered by many Jains in India and depicted in temple iconography (Glasenapp 1999 [1925]: p. 50). The term 'God' is part of popular Jain vocabulary and is often used when communicating with people of other faiths as a symbolic representation of omniscience or the principle of enlightenment. Sadhvi Shilapi's words, for instance, were addressed to a primarily western audience whose central concerns were the environment and development rather than theology, but who would be broadly familiar with theistic language.

While this explanation might seem to digress from the main argument, it tells us much about the Jain attitude of pluralistic acceptance while maintaining cultural integrity. The many-sided philosophy is reinforced by the need to survive as a minority population and preserve cultural distinctiveness and simultaneously integrating with and participating in the surrounding society. Furthermore, many Hindu deities have an ecological significance and are perceived as archetypes or symbols of abstract philosophical ideals. Saraswati, for example, is associated with learning and life-giving water. As such, she embodies the principles of pursuing knowledge and treating the natural world with respect. Revering such deities also helps to embed Jain communities in their local settings. In the Diaspora, and sometimes in India, Jains sometimes share communal facilities with Hindus. This contributes to the mistaken impression among the wider public (and among some Hindu nationalists) that Jainism is a subset or 'sect' of Hinduism.

The methods of Veerayatan are pragmatic (another reason for the absence of 'hang-ups' over theological niceties) and concerned with realistic objectives and results rather than ideological purity. There is no romantic concept of the 'pristine wilderness' of the kind that overshadows western environmentalism. Nor is there a Rousseauvian regard for peoples living close to nature and vulnerable to its ravages. The practical or 'realist' view of ecology epitomised by Veerayatan's work accepts the overriding value of biodiversity. As well as being inherently advantageous for human communities (in terms, for example, of food, medicine, agriculture, building materials and aesthetic enjoyment), biodiversity has a positive value to Jains from a spiritual perspective, reflecting the existence of jiva in all life forms and hence a common bond between them. Biodiversity also accords with the philosophical ideal of Anekant, because it represents a plurality of perspectives crossing the bounds of species and plant or animal classification. The *Tattvartha Sutra* of Umasvati makes clear that 'each soul [jiva] has its own view of the world, which differs from the view of other souls' (Tatia 1994: p. 10). That principle encompasses the 'souls' (jivas or units of conscious life) embedded in plants and animals as much as humans.

Loss of biodiversity is not only an environmental tragedy, but also part of the loss of knowledge brought about by careless actions. Such actions are usually linked to short-term economic expansion without heed to the consequences for future generations. The loss of plant species leads to a loss of indigenous or local knowledge ('plant wisdom') and an associated loss of opportunities for medical research. Knowledge that could have universal significance is lost with the impact of careless forms of 'development' on local and often ancient cultures and resources (see Beyer 2009; Dobkin de Rios 1992 for perspectives from Amazonia). The Jain concept of biodiversity is rooted in the two contrasting aspects of the Dharma: the intuitive, based on a feeling of intimate connection and continuity with the natural world, and the scientific or rationalist, based on experiment, evidence and the pursuit of knowledge. Poverty and ignorance are regarded as forms of *himsa*, destructive to those they affect and likely to result in further destructive activities visited on the environment. The development strategy of Veerayatan is far from non-interventionist. Nature can be worked with, adapted, improved upon. Non-interference, far from being an expression of Ahimsa, can sometimes become a direct cause of harm, leaving human communities, animals and plants vulnerable to catastrophes such as flooding and drought. In the same way, preserving local or indigenous cultures need not mean keeping them in a state of isolation, ignorance of the outside world (a version of *avidya*) and subject to backward cultural practices, such as the oppression of women and restriction of their educational opportunities.

Veerayatan's activities include building more resilient and safer homes less vulnerable to the elements, expanding academic and vocational education for adults and children (male and female), promoting the use of technology and encouraging commercial enterprises. All of these reforms are interconnected and while they increase human independence from nature, they also promote greater ecological awareness and the possibility of living more harmoniously with the environment, as opposed to the current state of human-induced hostile competition. Protection from the ravages of nature, along with better education, can encourage greater concern for conservation and animal welfare rather than the disconnectedness that accompanies uncontrolled urban expansion. Crucial to Veerayatan's work is a focus on the communities it serves being given the power to make decisions for themselves rather than becoming passive recipients of 'aid'. Connected to this is the idea of self-sufficiency, in keeping with Gandhi's principle of *swadeshi* and the *Sarvodaya* ('universal uplift') movement that developed from it after Gandhi's death, led by non-violence advocate Vinoba Bhave and influenced by Jain thought, in particular the example of Mahavira (Ostergaard 1985). This involves local production for local needs, with regional economies robust enough to enable their communities to survive culturally and avoid being absorbed into expanding cities or being used as cheap labour for corporations. The Jain-inspired work of Veerayatan and similar philanthropic organisations remains small-scale in the overall context of India's economic development. More orthodox models have, in general, prevailed, based on economic growth as an end in itself. Gandhi, after all, did not succeed in establishing *swadeshi* as a lasting economic model. The idea of localised economic self-sufficiency and cultural integrity has been displaced in turn by centralised planning and neoliberal capitalism, both rooted in western ideologies and working on the assumption that resources are limitless, at the disposal of humankind and that nature has to be 'conquered' rather than accommodated.

The work of Veerayatan provides a useful case study of Jain principles applied to modern economic and social problems throughout the Global South. More than that, it is a quiet challenge to the dominant ideologies of development, derived from Indic culture and based on the South Asian experience. It also challenges the romantic green sensibility cited above that tends to view poverty and simplicity as virtues, nature as an inherently benevolent force and modernisation as inherently corrupting. In other words, binary choices are avoided, as is the adversarial logic that underpins them. The unsentimental and experiential programmes favoured by this Jain-led organisation express in action the philosophy of multiple viewpoints: both/and rather than either/or. In this context, Veerayatan's intention is not to choose, in western terms, between modernity and tradition, education and folk wisdom, continuity and change, progress and the environment. All are viewed as valuable attributes that can reinforce each other in beneficial ways. At one level, respect for nature and its processes can be enhanced by education and greater material detachment from nature's excesses. At another, 'overdevelopment' through uncontrolled economic expansion can be resisted and the traditional strengths of local communities maintained, reinforced by appropriate use of technology along with professional and commercial activity. While successful at grassroots level, the methods of Veerayatan and similar organisations have as yet made only a small dent in the dominant ideologies and practices associated with development and economic growth. Yet they offer us a glimpse of what a non-western form of political ecology might look like, if we ascribe a new meaning to 'political' that encompasses compassionate action or, in Jain terms, *punya* or positive karma.

Philanthropic activities such as those of Veerayatan are concerned with positive material improvement as well as the mere alleviation of poverty. They might therefore be interpreted as contradicting the vow of Aparigraha, or non-possessiveness. Within Jain communities, there has always been a creative tension between the belief in renunciation or focus on the inner self, and the active, outward-looking emphasis on commercial and professional success accompanied by philanthropic acts. In essence this is a tension between positive karma and the cessation of karma, *punya* and Samvara. It is resolved in part through the two types of vow. The *Mahavratas*, or Greater Vows, point towards withdrawal from the world and the cessation of karmic influence.

68 *A radical synthesis*

The *Anuvratas*, or Lesser Vows, point towards constructive activities that reduce or preferably avoid harm. As is usually the case with Jain Dharma, there are many shades in between. The ascetics of Veerayatan are still engaged in worldly, and hence karmic, activities and so are not in the ultimate phase of renunciation. Similarly, laymen and laywomen, especially those of advanced age, might divest themselves of most of their possessions (usually to family members and charities) and cease to perform most activities. By so doing, they emulate as far as possible the experience of ascetics. There is also a belief that the various stages of release from karma are like rungs on a ladder, albeit with the possibility of falling backwards as well as progressing upwards. Under normal circumstances, *punya* leads to *samvara*, which in turn leads to *nirjara*. Positive actions create the spiritual framework for the blockage and falling away of karmic matter.

It follows from this idea of progression that one of the main purposes of material success, be it in business or the professions, is to benefit the community and the wider society. Aparigraha has an active and positive as well as a passive and renunciatory side. Far more than being only about non-possessiveness, it is an organised programme of giving back. It is not possible to have Aparigraha, in its true sense, without *Parigraha*, much as *Moksha* (liberation) cannot be attained without passing though *samsara*, the cycle of birth, death and rebirth from which liberation is sought. The prominence of *samsara* in Jain culture also leads to the assumption that working towards enlightenment can take many lifetimes. That produces a sense of perspective and long-term purpose along with the ethical imperative to avoid *himsa* and live according to the Five Vows. In the context of business, the intention is (in theory at least) greater than the profit motive. It is the enrichment of the whole Jain community served by the enterprise, in addition to society in general. For that reason, commercial activity should involve a commitment to 'giving back' through communal or philanthropic activities. The majority of Jain-owned businesses remain small- or medium-sized and family orientated. Cultural preservation partially explains this, but there are equally important ethical reasons. For the values of a commercial enterprise are viewed as contributors to its success, not as luxuries or optional extras. There is a sense in which every Jain business is held to have a social dimension by its very existence. This is rarely broadcast because it is viewed as self-evident. Importantly, there is considered to be a natural limit to the size of an enterprise. Its growth takes place within an environmental context and the need to expand is balanced against the harm that expansion might do to the ethics of the business and to the community.

Expansion is followed by consolidation and sometimes contraction, so that the underlying values of the business can be retained and to ensure that the wealth it generates is shared. A good example is the jewellery trade, which has deep roots in Jain culture. In Jaipur, Rajasthan, it survives as a means of transmitting values as well as traditional skills from one generation to the next, adapting to modern conditions but retaining its inner core.

In an unusually direct account of the Jain jewellery business, its history and culture, the Jaipur entrepreneur Jyoti Kothari, owner of Vardhaman Gems, described to *Jain Spirit* magazine the way in which the industry is still based on workshop-centred craftsmanship. Manufacturing skills are for the most part transmitted between father and son, in an industry that still remains the preserve largely of men. The consumers, by contrast, are mainly female and so women play an important role in the design and marketing of the jewellery produced by Vardhaman and similar Jain-led firms. In most cases, the apprentice jeweller is a grandson or another member of the extended family. Immersion in the ethos and values of the industry is an integral part of his training, as important as the practical aspects. Indeed, in contrast to western methods of training, distinctions between 'theory' and 'practice' are barely drawn. This stance reflects the Jain system of logic, whereby intention invariably determines outcome. The apprentice undergoes a process akin in many respects to a spiritual initiation. His work is a form of commitment to the *Anuvratas*. While maintaining this traditionalist stance, Vardhaman Gems and other firms in the sector have become full participants of the outward-looking India of the twenty-first century. Kothari also makes explicit connections between the jewellery trade and Jain ethics, which are important to an understanding of the intentions and priorities and of the business community as a whole:

> Many modern Jains face a quandary when attempting to combine their personal values with their career aspirations. Yet, the jewellery trade is, arguably, a business that has the power to corrupt its professionals due to the pressures that come with dealing with high-value items. An honest jeweller is a rarity, yet many Jains have found prosperity through the trade precisely because of their religious beliefs and strong reputations. As Jainism advises that achieving purity is determined through facing and overcoming life's temptations, I would argue that the core religious values fit well with the honour code of the jeweller.
>
> (Kothari 2004: pp. 48–50)

Practising Jains also regard the gemstone and jewellery trade as *alparambhi*, or 'requiring minimum violence' (Kothari 2004: pp. 48–50). This explains why the trade is so attractive to members of a culture that seeks to avoid inflicting *himsa* on the living Earth. In this context, the jewellery industry in this refers to the jeweller's craft and the skills of gem identification and grading, rather than the process of gemstone extraction, which is unlikely be undertaken by devout Jains. That said, the jeweller or gemstone grader's lack of violence against people or ecosystems has made these occupations 'ideal for Jains wishing to adhere to the principle of ahimsa (*sic*)':

> Motivation for wealth earned with morality (*nyaya sampanna vaibhava*) was exactly the solid background for ethics and morality that fit in with Jain values. The high character and moral conduct of the Jains enabled them to be trusted by kings and aristocrats. The love for one's religious community (*sadharmi-vatsalya*) also played its role, and the established jewellers contributed to this growth by training generations of Jains with the secrets of the trade, which led to many Jains prospering in the trade.
>
> (Kothari 2004: pp. 48–50)

The skill of the jeweller, including the identification, grading and valuation of gemstones, is transmitted from teacher to pupil in a relationship that transcends apprenticeship and resembles in many ways the relationship between guru and disciple:

> The teacher nurture(s) his students with the qualities required of a true jeweller, whilst equipping them with the necessary practical skills and theoretical knowledge of the trade. They were taught to be patient, calm, vigilant, creative and diplomatic, fitting the sort of values Jains were traditionally taught. Jain jewellers functioned according to the following basic rules in particular: imitations were never to be sold as real; substituting of goods was treated as a major offence; a certain percentage was deducted in every transaction for charitable purposes.
>
> (Kothari 2004: pp. 48–50)

These attitudes and practices confer on the jewellery trade the status of 'vocation' or 'most honoured profession', requiring careful thought and action from its practitioners. Careful thought includes avoiding the greed or lust for possessions that can be easily associated with gems and jewellery. The beauty and intricacy of the gemstones is valued instead. The tradition of scholarship and inquiry associated with the Jain educational ethos is also brought into play. During the fourteenth century CE, Thakur Theru, a court jeweller in Delhi, published the *Ratnapariksha* (Gem Inspection Manual) and in the introduction 'clearly highlights the importance of [the Jain] religion to him and his trade' (Kothari 2004: pp. 48–50). The link between jewellery and spiritual practice is made explicit in major texts such as the *Kalpasutra* (first century CE), where gem identification is listed as one of seventy-two principal skills to be acquired by men. Trishala, the mother of Mahavira, is also quoted as describing heaps of gemstones appearing in the thirteenth of fourteen auspicious dreams before her son's birth.

Significantly, Jains seek to live according to the Three Jewels or 'Triple Gems' (*Ratnatraya* or *Triratna*) cited briefly above. These might also conveniently be referred to as the three *Samyaks*, literally three 'rights', signifying compatibility with Dharma as guiding principle of the universe. The first of these, *Samyak Darshana* (Right Vision or Viewpoint), is associated with clarity of perception. *Samyak Gyana* (Right Knowledge) is associated with sifting fact from fiction and truth from lie in the same way that a jeweller separates gem quality from opaque stones or distinguishes between real and synthetic gems. *Samyak Charitra* (Right Conduct) is based on making the informed choice to act carefully and ethically, in as close as possible alignment with Dharma. The image of the jewel is also applied to the concept of Anekantavada (many-sidedness or). Each *naya* or viewpoint is likened to one of the facets of a cut diamond so that the same clear light is viewed from many angles. It is no coincidence that Vardhaman Gems is named in honour of Vardhamana Mahavira.

Through the jewellery trade, Jains have expressed their adaptability and internationalism. In India, they have served as court jewellers for Hindu and Muslim rulers alike, their influence surviving and growing under the British colonial administration. During the twentieth century, Jain families involved in the jewellery trade moved to areas as distant as Thailand and Japan. In Kobe, Japan, they formed a community associated with the pearl industry. Today, Jains continue to work as jewellers, gemmologists and diamond cutters in the principal markets of New York, Antwerp, Geneva and London. East Africa attracted Jain jewellers from Gujarat in the late nineteenth and early twentieth centuries because of the opportunities afforded by the discovery of gold and diamond deposits. They continue to play an important role in the Diaspora communities of the region, as well as in the gold and diamond industries of southern Africa. Charitable giving is important to the lives of Jain families in the jewellery trade:

> A large number of educational institutes, hospitals, rehabilitation centres, *dharamshalas* [religious guest houses; lodges for pilgrims] and animal welfare centres are run by Jains in various parts of India. Many prominent charitable trusts were established by the Jain jewellers for social and religious services, and it is the jewellers' contributions that have maintained some of the Jain temples and *upashrayas* [lodging houses for ascetics] built outside of India ... [and] the conservation of old temples in India.
>
> (Kothari 2004: pp. 48–50)

This long-established industry serves as a template for other Jain businesses. In one sense, Vardhaman and similar companies in the trade (often more like colleagues than rivals) are expansive, internationally minded and receptive to change. In other ways, they remain largely family-centred and focused predominantly on the surrounding Jain population. Vardhaman is culturally conservative, in that it perpetuates practices that have been successful over many generations and retains an extensive network of social obligations and responsibilities. Much of Jain commercial activity remains community based and cleaves to its local or familial roots during the process of expansion. Indeed, as the enterprise expands, those roots usually become longer and deeper. Yet, for a company like Vardhaman, the limits of expansion are set when it becomes incompatible with its core values and practices and loses touch with its communal base. The firm's economic activity is a continual balancing act between its outward-looking, cosmopolitan attitude to trade and retaining its cultural anchor in the Jain business networks of Jaipur. Centrifugal and centripetal principles complement each other, with charitable donations and traditional systems of apprenticeship being at least as important as international recognition. In other words, it is accepted by a company like Vardhaman Gems that there are natural limits to growth for corporations much as there are for living organisms.

At the same time, there is no contradiction between short-term commercial success and a larger, long-term concern for the environment and community. The two positions are not viewed as polar opposites, as is so often the case in western political discourse, but as points on a continuum. They are also principles that reinforce each other. The position adopted by Vardhaman Gems is typical of Jain businesses in India and the Diaspora. It is founded on accumulated experience as well as an ethical basis derived from the vows undertaken by laymen and laywomen. Ownership of a commercial enterprise is seen in terms of obligation to others. It confers responsibilities towards extended family members, the community and society, including the environment and the natural world. In other words, ownership is a form of tutelage of the business, which is held in trust for future generations, to whom it also 'belongs'. An aspect of this stewardship is the avoidance or reduction as far as possible of practices that could cause harm, whether to people, animals, plants or the atmosphere. This is partly because such harm is seen as putting the next generation at risk, and with it the future of the business and perhaps the whole Jain community. Yet it is equally rooted in the sense that all life forms (within the *Lokakasa* or 'inhabited universe') depend on each other for their survival.

The strong connections made between present actions and future outcomes, as well as between ownership and moral responsibility, derive from the emphasis of Jain thought on interdependence and long-term preparation. 'Success' is defined in terms of stewardship and careful action as well as profit margins. The broadly 'pro-business' attitude takes the form of support for independent enterprises, generally small, medium-sized and family based. These businesses are viewed as an important countervailing force to corporations with impersonal, bureaucratic structures. In Jain commercial ethics, there is a presumption in favour of consolidation rather than expansion. Expansion can take place for pragmatic reasons, such as preserving an important aspect of the business or to serve Jain communities in other parts of the world. The principles and integrity of the business need to be kept in mind. A connection is therefore made between preserving the size of the enterprise – and its familial or communal roots – and preserving its values. The concept of Aparigraha or non-possessiveness (in a business setting the most important vow apart from Ahimsa) means that personal ownership is regarded as by definition impermanent and that there are natural limits to the size of an enterprise. Expansion beyond those limits has a detrimental effect internally (on the character of the business) and externally (on its relationship with the environment and society). Thus a commercial enterprise, however family-centred and traditionalist it might appear, is not an isolated entity but part of something larger, with an obligation to 'give back' to the surrounding community. In Jain cosmology, the balance between continuity and change is central to an understanding of how the universe works. That principle is transferred to managerial ethics: excessive conservatism tends to lead to stagnation, whereas a constant state of flux creates instability and the loss of values.

Jain businesses rarely describe themselves as social enterprises. They do not need to do so, because there is no concept of a business that it anything other than 'social' in their wider cultural setting. It is in the commercial sphere, which has always been of great importance to Jain communities, that a distinctive yet low-key political consciousness is most fully expressed. It is not an ideology in the formal sense. In many ways, it is better described as a disposition, akin to the sensibility of 'Jainness', which is pluralist in spirit and makes connections rather than compartmentalising or building walls. *Jiva Daya*, as a form of imaginative and sympathetic identification with all sentient beings, is a moral sensibility that acts as a restraining influence on human activity. The Jain political disposition is founded on these intuitive conditions, combined with a profound belief in the doctrine of cause and effect, expressed through the imagery of karmic accumulation. Every action has direct consequences for the individual (positive or negative karma, spiritual progress or regression) and his or her surroundings, while choices made in the present affect future generations, often in unpredictable ways. All action is karmic, even the most altruistic or creative, and so it must be pursued with meticulous care. Intellectual rigour does not mean adopting fixed or doctrinaire positions. Instead, it implies the ability to reflect on multiple viewpoints; this principle is extended to commercial decision-making, which is not considered separate from the life of the mind.

Many of these conclusions have equivalents in political ecology and even 'soft' environmentalism. The concept of Careful Action corresponds in many ways to that of the 'ecological footprint' that is an expression of the planetary impact of an organism or species, including human individuals and civilisations (Chambers, Simmons and Wackernagel 2000). Reducing the size and depth of our individual and collective ecological footprint is an ambition cited frequently by political ecologists and environmental campaigners. Aparigraha, practised through the consolidation of a business rather than its limitless expansion, can be said to resemble the western idea of 'limits to growth' (Meadows and Meadows 1972; Meadows and Randers 2004). Green politicians and campaigners often speak of 'holistic' approaches to the environment, economics and the political system. Applied in this way, the concept of holism based on 'linkage and reciprocity' rather than 'splitting things up' (Dobson 2007: p. 30; Smuts 2013 [1926]) has much in common with the connections Jains make between self-interest and altruism. Yet these correspondences are not precise and Jain thought should not be assumed to echo or mirror political ecology. The distinctive position of Jain communities towards the integration of business and the environment challenges many of the centralising assumptions of western green movements, which tend to place a far higher emphasis on the role of the state as the instrument of change. This is because the Jain method of conducting business along ecological lines is shaped by different cultural and political experiences from those of western greens. It has evolved independently of European and North American thought but does not dogmatically exclude western philosophical and scientific influences.

Although the way of life from which it emerged remains cautious and conservative in its outward expressions, Jain philosophy is radical in the true sense. All received wisdoms, including those of political ecology, are called into question and examined from the roots upwards. In this way, the Dharma of this numerically small but quietly influential section of Indian society can potentially affect or even reframe the way environmentalists and ecologists think about politics. At the time of writing, there is no directly 'Jain-based' form of environmental politics or green philosophy. The contribution has been indirect and subtle. Jains, as individuals, families and businesses, participate in environmental movements and pressure groups. In so doing, they are influenced by their culture's traditional emphases on the interdependence of all living organisms, the moral responsibility of human beings to act with restraint and, beyond that, the spiritual ideal of non-possessiveness. Sometimes, young professional Jains rediscover their cultural roots through environmentalism. Organisations such as Young Jains in the United Kingdom, the Jain Association of North America (JAINA) in the United States and Canada, along with Project Anveshan in Bengaluru (Bangalore), India have a strong ecological component in their propaganda and activities (Shah and Rankin 2017: pp. 86–7, 127). Their principal appeal is to university students and graduates working in the academic, professional or commercial spheres, who are products of an urban, secular and pluralist education.

For many of this younger generation of Jains, awareness of atmospheric pollution, climate change and inequalities at local and global level have rekindled an interest in the philosophy that had served as a backdrop to their lives. Like previous generations, they are interpreting Jain Dharma in a way that seems relevant to their lives and the concerns of the society around them. Environmental issues enjoy a pre-eminent status and the tradition of strict vegetarianism is emphasised as an expression of Ahimsa (Shah and Rankin 2017: pp. 86–7). This practice, along the austerities followed by a small of minority of ascetic men and women, has at times caught the imagination non-Jain environmentalists and greens, chiefly in western societies (see Chapter 2).

They observe these forms of abstinence and see them as a possible model for an ecological lifestyle. In an age of environmental activism, with a mounting – and evidence-based – sense of crisis, concepts such as Ahimsa (avoidance of harm) and *alparambhi* (work that 'requires minimum harm') take on a new urgency. The overproduction of meat through intensive farming, the pressure of rising human populations on the habitats of other species and the increasingly unequal distribution of food, water and other natural resources all necessitate changes in our patterns of living. More than that, they require a rethinking of our expectations and the way we use our intelligence and creative powers. It is here that the Jain philosophical tradition, with its long-term perspective and constant interrogation of human thoughts, motives and desires, can be of greatest benefit.

Bibliography

Bahro, R. (1982) *Socialism and Survival*, London: Heretic Books.
Benton, T. (1993) *Natural Relations: Ecology, Animal Rights and Social Justice*, London: Verso.
Benton, T., ed. (1996) *The Greening of Marxism*, New York: Guilford Press.
Beyer, S.V. (2009) *Singing to the Plants: A Guide to Mestizo Shamanism in the Upper Amazon*, Albuquerque: University of New Mexico Press.
Chambers, N., Simmons, C. and Wackernagel, M. (2000) *Sharing Nature's Interest: Ecological Footprints as an Indicator of Sustainability*, London/Stirling: Earthscan.
Dobkin de Rios, M. (1992) *Amazon Healer: The Life and Times of an Urban Shaman*, Bridport: Prism Press.
Dobson, A. (2007) *Green Political Thought*, 4th edn, Abingdon: Routledge.
Glasenapp, H. von (1999) [1925] *Jainism: An Indian Religion of Salvation*, New Delhi: Motilal Banarsidass.
Jain, J. (1993) 'Local Customs Recognised by Jains', *Journal of the Asiatic Society of Bombay*, vols 64–66, pp. 85–94.
Jaini, P.S. (2001) *The Jaina Path of Purification*, 4th edn, New Delhi: Motilal Banarsidass.
Kothari, J. (2004) 'A Diamond is Forever', *Jain Spirit*, Issue 20 (September – November 2004), pp. 48–50.
Kropotkin, P. (2014) [1902] *Mutual Aid: A Factor in Evolution*, Scotts Valley, CA: CreateSpace Independent Publishing Platform.
Mardia, K.V. (2007) *The Scientific Foundations of Jainism*, 4th edn, New Delhi: Motilal Banarsidass.
Mardia, K.V. and Rankin, A. (2013) *Living Jainism: An Ethical Science*, Winchester: Mantra Books.
Meadows, D. and Meadows, D. (1972) *The Limits to Growth*, New York: Signet.

Meadows, D. and Randers, J. (2004) *Limits to Growth: The 30-Year Update*, Hartford, VT: Chelsea Green Publishing Co.
Morris, W. (1993) *News from Nowhere and Other Writings*, Harmondsworth: Penguin Classics. [*News from Nowhere* was originally published in 1890.]
Ostergaard, G. (1985) *Nonviolent Revolution in India*, New Delhi: Gandhi Peace Foundation.
Rankin, A. (2010) *Many-Sided Wisdom: A New Politics of the Spirit*, Winchester: Mantra Books.
Shah, A. and Rankin, A. (2017) *Jainism and Ethical Finance: A Timeless Business Model*, Abingdon: Routledge.
Shilapi, S. (2002) 'The Environmental and Ecological Teachings of Tīrthankara Mahāvīra', in Chapple, C.K., ed. *Jainism and Ecology: Nonviolence in the Web of Life*, Cambridge, MA: Harvard University Press, pp. 159–169.
Smuts, J.C. (2013) [1926] *Holism and Evolution*, Gouldsboro, ME: Gestalt Journal Press.
Tatia, N. (1994) *That Which Is: Tattvārtha Sūtra (Umāsvāti)*, San Francisco, CA: HarperCollins.
Zimmer, H. (1969) *Philosophies of India*, new edn, Princeton, NJ: Princeton University Press.

Index

Adipurana (ninth-century CE text) 64
agricultural methods 6–7, 35, 65, 76
Ahimsa (non-violence in action, speech, and thought): and accumulation of wealth 25, 30, 39, 70; and Anekant (many-sidedness) 4, 6, 28, 39; and Aparigraha (non-possessiveness) 26, 30; and business community 7, 25, 70, 73; and Careful Action (*Irya-Samiti*) 6, 18, 25; centrality of in Jain doctrine 1, 3, 22, 24, 25–6, 39, 42, 54, 62; and environmental politics 32, 76; and exercise of tolerance 20; and Gandhi 7, 33–4, 46–7; as power conferred by human intelligence 57; and Veerayatan movement 63, 66; and younger generation of Jains 75
ajiva (inert or 'soulless' matter) 12, 15, 32, 54, 55, 56
Ajnanavada (scepticism) 41–2
Akriyavada (non-action) 42
alparambhi ('requiring minimum violence') 70, 76
Anekantavada nor Anekant (doctrine of many-sidedness) xii; and adversarial mode of politics 8, 20, 47, 54; and Ahimsa 4, 6, 28, 39; and western thought 38–40, 45, 46, 47–50, 51; centrality of tolerance 4, 8, 20, 38, 40, 64–5; and commercial decision-making 31, 51, 74; cut diamond analogy 13, 37–8, 71; as a defining concept of Jain Dharma 38–40, 45; doctrine of both/and 21, 34, 40, 51, 67; as environmental principle 6, 9, 19, 35, 44–5, 46, 47–50, 51; and freedom of conscience 62; as inherently pluralist concept 6, 9, 19, 27, 54, 64–5; and Jain logic 21, 34, 35, 37–9, 40–5, 47–9, 51, 67; and *naya* of other species 19, 39, 45, 59, 65; philosophy of 'perhaps' and 'maybe' 40, 43; and reality/existence 39–45, 54; relationship with *Syadvada* system of logic 43–5; as survival mechanism for Jain communities 48, 64–5
Anekant Diamond Products, Jaipur 37
animal welfare 2, 9, 60, 61, 63, 66, 73; as equivalent to human welfare 28, 54, 62; *Panjrapoor* (animal sanctuaries) 3, 22, 62, 72; and *punya* 14, 33, 55
anityavada ('non-eternalism') 43
Antwerp, Belgium 21, 71
Anuvratas (Lesser Vows) 25, 26, 50–1, 62, 67–8, 69
Aparigraha (non-possessiveness) 25, 26, 30, 39, 40, 50–1; and business community 25–6, 30–1, 33, 67, 68–9, 70–1, 73, 74, 75; and giving back 3, 7, 22, 26, 30, 39, 61, 68, 72 (*see also* philanthropic tradition); and green political thought 29–31; and Jain environmental thought 29–31, 51, 58–9, 73, 74; as path towards *Nirjara* 58–9
artistic creativity 19, 22, 27, 28

ascetics: aim of escape from natural world 3, 5, 15, 17, 26, 49, 50, 51, 56, 61; austere practices of 1, 2, 5, 19, 24, 26, 30, 44, 50, 56, 75–6; commercial support for 3; doctrine of 'self-conquest' 3, 5, 60; female 6–7, 18, 33, 50, 62–3, 68; *Irya-Samiti* (Careful Action), vow of 18; as Jain ideal 15, 17, 50; and *Mahavratas* 25, 26, 30, 50, 62; as distinct from 'monks' or 'nuns,' 8–9; and two 'schools' of Jain practice 44; and western greens 5, 6, 19, 34, 75–6; male 1, 33, 44
asrava (inflow of karma) 55, 56
Asteya ('Non-stealing,' avoidance of theft or exploitation) 25, 26
astikya (implicit understanding of the nature of reality) 57
Avasarpini (regressive half-cycle of time and the universe) 27
avidya (ignorance, lack of knowledge), cycle of 13, 17, 20, 56, 66

bandha (karmic bondage) 56
Bhave, Vinoba 66
Bihar, northern India 6–7, 33, 50, 62–3
biodiversity 9, 65–6; loss/extinction xii, 65
Brahmacharya (chastity, fidelity and avoidance of promiscuity) 25
Buddhism 12, 14, 24, 33, 43, 64; Theravada 43; Zen 47
business community: co-operative enterprises 33, 59, 60; as culturally conservative 72; and diaspora 3, 21, 71–2; Jain ethos 3, 7–8, 15, 19, 30–1, 33, 51, 56, 57, 68–73, 74; and *Jiva Daya* 74; limits to the size of enterprises 31, 33, 68–9, 72, 73, 74; and philanthropic tradition 3, 7, 60, 61, 68, 72; practice of Aparigraha 25–6, 30–1, 33, 67, 68–9, 70–1, 73, 74, 75; and SMEs 7, 33, 68, 73; success as highly valued 3, 15, 18, 19, 20–1, 22, 23, 31, 67, 68, 72–3; sustainable enterprise 18–19, 33, 47, 60; traditional systems of apprenticeship 19, 69, 70, 72

California (Green Party in) 45–6, 47–8
Campaign for Political Ecology 8
charvaka ('annihilationism') 42
Christianity 4, 9, 43; concept of soul 16
civil rights movement in West 46
'civilian' Jains xii, 1, 2–3, 5, 17–18, 49–50, 60, 68; and *Anuvratas* 25–6, 27, 30
climate change xii, 63, 75
cosmology, Jain: balancing of continuity and change 7, 22, 27, 31, 32–3, 38, 41, 53, 61, 73; cycles of the universe 15, 22, 27, 28, 40, 41, 64; cyclical conception of time 3, 22, 40, 41, 64; Dharma (cosmic order) 21–2, 39–40, 53–4, 56; emphasis on multiplicity rather than duality 41–2, 44; journey of the individual soul 3, 6, 13–18, 19, 22–3, 24–6, 37–8; long half-cycles 27, 28, 41, 64; long-term view 3, 5, 15, 22, 61, 68, 72, 73; perception of matter 12, 14, 54–5, 56; transcendence/self-liberation/salvation 3, 5–6, 13–19, 22–3, 24–6, 31, 38, 49, 51, 56, 59, 61; *see also* karma, Jain concept of
counter-culture (1960s) 46, 47

Daoist thought 41
Darshana (perception, clarity of vision) 57
Deep Ecology platform 8, 48–50
deforestation 35
Devall, Bill 8
Devanagari script 28
Dharma (cosmic order) 21–2, 39–40, 53–4, 56
diaspora, Jain 19, 21, 29, 44, 48, 75; and Hinduism 27, 64; and jewellery trade 71–2
Digambar ('Sky-Clad') ascetic tradition 44
dogmatism 6, 12, 20, 54, 60–1
dravya (substance) 43
dualism 41

ecological footprint (concept of) 6, 18, 74
Ecosophy T (theory of Arne Naess) 8
education 3, 21, 28, 35, 39, 60, 75; Jainism's high regard for 5, 7, 15, 19–20, 26, 33, 58, 62, 71; and Veerayatan movement 33, 50, 63, 66, 67; of women 33, 58, 66
Ekantika (*Ekant*) (one-sided viewpoint) 20, 39, 42, 54, 57, 58
ekatva (persisting unity) 27, 40
Enlightenment, European 34, 35, 48
environmental crisis xii, 5, 9, 76
environmental politics; conservative sub-theme 33; defined 4, 8; eco-socialists 49, 59–60; environmentalism 8, 15, 19, 50, 65, 74, 75; Green Party of California 45–6, 47–8; Nonviolent Direct Action (NVDA) 32, 33, 34; political ecology or ecologism 8, 18–19, 31, 34, 45–51, 59–60, 62–8, 74–5; prevailing adversarial mode 19, 47, 54, 67; problems of terminology 8; urban base of European Greens 46, 49; and Veerayatan movement 62–4; view of scientific and technological change 34–5, 46, 59–60; *see also* green thought, western
environmental thought, Jain; aim of escape from natural world 3, 5–6, 22–3, 35, 49, 56, 61; and Anekant 38–9, 40, 44–5, 47, 49–51, 64–5, 74; and Aparigraha (non-possessiveness) 25–6, 29–31, 51, 58–9, 73, 74; 'Careful Action' principle 6, 8, 18, 25, 29, 31, 42, 44–5, 73, 74; concept of biodiversity 65–6; embedded into lives of 'civilian' Jains 1, 2–3, 5, 17–18, 24, 26, 62, 75–6; importance of *avidya* 20; as long-term view 3, 5, 15, 22, 61, 68, 72, 73; and Mardia's concept of 'Jainness' as a sensibility 22, 27–8, 31, 74; nature as not romanticised 6–7, 35, 49–50, 65, 67; and prevailing model of economic growth 31–2; *Samyak Darshana* disposition 57–8; sustainable enterprise 18–19, 33, 47, 60; western green thought as distinct from xiii, 2, 6, 15, 29–35, 49–50, 61–2, 74–5; *see also* Veerayatan movement

Gandhi, Mahatma 7, 33–4, 46–7, 60, 66
Gautama Buddha 24, 33
gender: and California's Greens 45–6; ecofeminism 49; economic independence of Jain women 33; female ascetics 6–7, 18, 33, 50, 62–3, 68; female education 33, 58, 66; female literacy rates 19; and jewellery trade 69; male ascetics 1, 33, 44; and Veerayatan movement 66; Global South 5, 7, 67
globalisation 39, 48
Green Party of California 45–6, 47–8
green thought, western: and alternative business models 7–8, 59; common ground with Jain thought 29, 31, 32–3, 46, 47–8, 50; 'eco-warrior'/activist mode 2, 7, 19, 29, 46; embrace of left-wing ideologies 8, 34, 47, 49, 59–60; green sensibility 34–5; and Jain asceticism 5, 6, 19, 75–6; Jain thought as distinct from xiii, 2, 6, 15, 29–35, 49–50, 61–2, 74–5; partisan ideology of 29, 31, 33, 34–5, 49, 54; as response to industrialism/mass consumption 29, 34–5, 48–9; and role of state 8, 49, 75; romanticisation/idealisation of nature by 6, 48–50, 65, 67; Rousseauvian 65; urban base of European Greens 46, 49; *see also* environmental politics

Haudenosaunee (Iroquois) people 5
himsa (violence, harm, destructive power) 25, 26, 58, 66, 70
Hinduism 2, 4, 12, 14, 21, 23, 27, 33–4, 55, 64–5
homo economicus ('economic man') 35

India 2, 3, 5, 6–7, 15, 19, 27, 48, 64; Gandhi's *satyagraha* ('truth-struggle') 34, 46; Gandhi's *swadeshi* (economic self-sufficiency) 33–4, 46–7, 60, 66; post-independence 'development', 60, 66–7; *Sarvodaya* ('universal uplift' or 'compassion for all') movement 66
individuality: as central to Jain theory and practice 18, 22–3, 24, 33, 35, 53–4, 61; doctrine of 'self-conquest' 3, 5, 24, 56, 60; dual concept of 14, 53–4; and liberation/salvation 3, 13–18, 19, 22–3, 24–6, 49, 59, 61
inequality 46, 49, 60, 75, 76
Irya-Samiti (principle of Careful Action 6, 8, 18, 25, 29, 31, 42, 44–5, 70–1, 73, 74

Jai Jinendra (traditional Jain greeting) 24
Jain Association of North America (JAINA) 75
Jain communities: and accumulation of wealth 3, 5, 15, 20–1, 25, 29–31, 39, 70; cultural distinctiveness of 4, 31, 45, 48, 62, 64–5, 74–5; ethos of self-reliance 3–4; high regard for education/knowledge 5, 7, 15, 19–20, 26, 33, 58, 62, 71; as perpetual minority 3–4, 21, 38, 48, 64–5; personal achievement and material success 3, 5, 18, 19, 20–1, 23, 30, 56, 68; as predominantly urban 20–1; seniority in 30, 68; social conservatism of 19, 31, 32–3, 61, 72, 75
Jain Dharma (Jainism): as ancient faith 1–2; Anekant as a defining concept 38–40, 45; apparent contradictions 7, 21, 42–3, 53–4, 67, 72; as based on continual questioning 2, 4, 6, 19, 29, 34, 40, 44, 47, 48; concept of soul 16, 28; as holistic in character 28–9, 34, 57, 74; intellectual modesty 44, 45, 54, 57; as intensely pragmatic 18–19, 24, 26, 44, 65–7, 74; lack of factionalism 20, 44; as non-theistic system 2, 23, 27, 64; openness to viewpoints from other cultures 2, 4, 6, 21, 27, 29, 59, 64–5; principle of restraint 1, 4, 5, 18, 51, 60, 75 (*see also* Ahimsa (non-violence in action, speech, and thought)); pro-technology stance 4, 7, 32, 33, 34, 35, 50, 60, 66; as radical in the literal sense 53–4, 75; rational and intuitive aspects of 2, 22, 27–8, 57–8, 65–6, 74; soft power and 'hard' tolerance 4, 21; terminology 8–9, 21–2; two 'schools' of 44
Jaini, Padmanabh 44
Jainness, sensibility of 22, 27–8, 31, 58, 74
Jaipur, India 7, 19, 37, 69, 72
jewellery trade 19, 21, 37, 69–73
jina, status of 23, 24, 42
jiva (life monad, animating principle): and biodiversity 65; contact with karma 13, 14, 15–18, 43, 54–5, 56, 58–9, 64; Jain view of 55; *leysha* (karmic colour) 11, 15–18, 58; liberation/return to essence 14–15, 23, 24, 41, 53, 56; 'modes' of existence 14, 17, 27, 41, 43–4, 53, 54–5, 56, 58–9, 68; as permanent 27, 28, 41, 43; possessed by every life form 12, 13, 19, 25, 31, 59, 65; possibility of regression 14, 27, 56, 58, 61, 64, 68;
Jiva Daya (sympathy or identification with all sentient beings) 20, 28, 58, 74

Kāla (time) 64
Kalpasutra (first century CE text) 71
karma, Jain concept of: cause and effect doctrine 12, 74; as distinct from other Indic traditions 5, 12; illusory attachments 3, 54, 56–7; karmic colour theory 11–12, 15–18, 58; karmic entanglement/attachments 3, 5, 6, 13–18, 23, 40, 42, 43, 49, 54, 55–7, 58, 64
karmic particles (karmic matter, *karma pudgala*) 12, 13–14, 15–18, 20, 23, 24, 25, 26, 40, 54–7, 58–9, 68; 'karmons' 54–5; literal and metaphorical views of 15; and *Mithyatva* (false belief or false consciousness) 58; and *Mohaniya* (self delusion) 56–7; *papa* (negative

or destructive karma) 14, 16, 55, 58–9, 60–1; *punya* (positive karma) 14, 16, 26, 32, 33, 55, 58–9, 60, 62, 67, 68; *samvara* (stoppage of karmic influx) 56, 67, 68; variants of karmic matter 14, 54
kasaya (negative passions) 24, 54, 58
knowledge: absolute truth 37–8, 42, 57, 61; and biodiversity loss 65; concept of humility 6, 20, 57; cut diamond analogy 13, 37–8, 58, 71; identified with light 13, 16–17, 20, 37–8; ignorance as form of *himsa* 66; incompleteness of 1, 40–1, 43, 57, 61; inner self as source of 18; Jainism's high regard for 5, 7, 15, 19–20, 26, 33, 58, 62, 71; and jiva's spiritual journey 53, 54–7, 58–9; Three Jewels (*Ratnatraya* or *Triratna*) 37, 57–8, 71; wealth as analogous to 30; western compartmentalising of 28–9, 54
Kobe (Japan) 71
Kothari, Jyoti 69–71
Kropotkin, Peter 59–60

legal profession (Jains in) 3, 21
leysha (karmic colour[s]) 11–12, 15–18
liberalism, western 22
life forms: avoidance of injury/harm to (*see* Ahimsa (non-violence in action, speech, and thought)); human capacity for delusion 1, 20, 56, 57, 58; humans as having highest number of 'senses,' 16–17, 56–7; humans as most spiritually evolved 14; jiva as possessed by all 12, 13, 19, 25, 31, 59, 65; life principle 2; minuscule/elementary 1, 4, 12–13, 18, 26; mutual dependence/interconnectedness 2, 5, 12, 13, 18, 27–8, 31, 41, 49, 62, 73, 75; plants 25, 26, 65; purpose, identity and viewpoint 2, 3, 4, 6, 12–13, 26, 28, 38, 45, 48, 51, 59; *see also* non-human species
logic, Jain system of *see* Syadvada (Jain system of logic)
Lokakasa ('inhabited universe') 12, 73

Mahavira, Vardhamana (twenty-fourth *Tirthankara*) 23–4, 28, 33, 63, 66, 71
Mahavratas (Greater Vows) 25, 50, 62, 67–8
Mardia, Kanti 22
Maya (deceit or illusion) 58
medical profession (Jains in) 3, 21
meditation and reflection 6, 12, 19, 21, 22, 23, 39, 57
Mithyatva (false belief or false consciousness) 58
Mohaniya (karma leading to self-delusion) 56–7
Moksha (spiritual liberation) 14–15, 19, 35, 53, 54, 56, 57, 68
Morris, William 59–60
multi-cultural societies 39

Naess, Arne 8, 48, 49
nationalism, resurgent 39
natural disasters 6–7, 50, 63, 66
naya (viewpoint) 19, 29, 39, 43, 44, 45, 59, 65, 71
neoliberal capitalism 67
Nirjara (falling away of karma/karmic particles) 56, 58–9, 68
nirvana (full enlightenment) 43
nityavada (eternalism) 42–3
niyativada (fatalism) 42
Non-Governmental Organisations (NGOs) 62
non-human species: biodiversity loss xii, 65; complexity of 2; and concept of 'social' 1, 37, 56–7; minuscule life forms 1, 4, 12–13, 18, 26; respect for *naya* of 19, 39, 45, 59, 65; 'senses' or level of consciousness 16–17, 25
Nonviolent Direct Action (NVDA) 32, 33, 34

oceanography 13
optometrists (Jains as) 21

painting and sculpture 22
Panjrapoor (animal hospitals and sanctuaries) 3, 22, 62, 72
papa (negative or destructive karma) 14, 16, 55, 58–9, 60–1

84 Index

Parasparopagraho Jivanam (Sanskrit verse) 28, 53, 59
Parigraha (acquisitive thinking and behaviour) 54, 58, 68
Parshva (twenty-third *Tirthankara*) 23, 24
philanthropic tradition: as Ahimsa 25–6; as aspect of Dharma 22; and business community 3, 7, 60, 61, 68, 72; educational opportunities as integral 3, 7, 33, 50, 62, 63, 66, 67; and female ascetics 18, 33, 62–3; and goal of self-liberation 15; and karmic activities 14, 68; lack of ideological strings 61–2; and practical environmentalism 6–7, 33, 50, 62–3, 64, 65–7; practice of Aparigraha 25–6, 30, 39, 40, 67, 68; *Samyak Darshana* disposition 57; social pressure towards charitable giving 3
plankton 13
pluralism or multiple viewpoints *see* Anekant (many-sidedness)
political ecology or ecologism 8, 18–19, 31, 34, 45–51, 59–60, 62–8, 74–5
pollution 63, 75
poverty and underdevelopment 50–1, 60, 62, 63, 66–7
Project Anveshan (Bengalaru/ Bangalore, India) 75
punya (positive karma) 14, 16, 26, 32, 33, 55, 58–9, 60, 62, 67, 68

Rajchandra, Shrimad 33–4, 46
reality, Jain view of: and Anekant 6, 20, 39–45; *astikya* (implicit understanding of the nature of reality) 57; cyclical conception of time 41; degrees or levels of 27, 40, 41, 54, 55, 56, 59; *Samyak Gyana* or *Jnana* (Right Knowledge or Understanding) 37, 57–8, 71; *Shraddha* ('educated faith') 57–8; tattva ('Nine Reals') 41, 55–6; Three Jewels (*Ratnatraya* or *Triratna*) 37, 57–8, 71; universal or absolute truth 37–8, 42, 57, 61
Rishava (or Rishabha, first Tirthankara) 23, 64

samsara, cycle of 3, 5–6, 13–18, 19, 22–3, 42, 43, 55, 56, 59, 68
samvara (stoppage of karmic influx) 56, 67, 68
Samyak Darshana (Right Vision or Viewpoint) 37, 57–8, 71
Samyak Gyana or *Jnana* (Right Knowledge or Understanding) 37, 58, 71
Samyaktva 15, 30
sangha (Jain community) 2–3, 33
Saraswati or Sarasvati (goddess of wisdom and the arts) 64, 65
Sarvodaya, Gandhi's philosophy of 66
Satya (Truthfulness, honesty, personal integrity) 24
science 2, 5, 12–13, 20, 27, 34, 66
Sessions, George 48, 49
Shilapi, Sadhvi 63–4
Shiva (Hindu deity) 23, 64
Shraddha ('educated faith') 57–8
siddha (liberated jiva/'soul') 23
small- and medium-sized enterprises (SMEs) 7, 33, 68, 73
Smayak Charitra (Right Action or Conduct) 37, 58, 71
social engagement 21, 24, 26, 28–9, 37, 51, 61–2, 68: flexible definition of 60; and loosening of karmic 'chains' 56–7, 59; *see also* business community; Veerayatan movement
social justice/welfare xii, 1, 33, 46: extended to non-human species 1, 6, 37, 56–7, 60
socialist and anarchist traditions (western) 8, 34, 47, 49, 59–60
soil erosion 35
Sutrakrita (second *Anga*) 41–2
Svetambar (White Clad) ascetic tradition 44
Syadvada (System of logic): and Anekant 21, 34, 35, 37–9, 40–5, 47–9, 51, 67; based on both/ and 21, 34, 40, 51; as 'holistic' or 'integrative' 34; intention as invariably determining outcome 69; philosophy of 'perhaps' and 'maybe,' 40, 43; premise of partial or qualified definition 43–4; questioning of dogmatic certainty 12; and western-derived ideas 29

tattva(s) ('Nine Reals') 41, 55–6
Tattvartha Sutra of of Umasvati 28
technological progress 4, 7, 32, 33–5, 46, 47, 50, 60, 66, 67
Theravada Buddhism 43
Theru, Thakur, author of *Ratnapariksha* (Gem Inspection Manual) 71
Three Jewels (*Ratnatraya* or *Triratna*) 37, 57–8, 71
Tirthankara(s) (ford-maker[s], pathfinder[s] or source[s] of spiritual inspiration) 22, 23–4, 27, 57, 62, 64
tolerance 4, 8, 20, 38, 40, 64–5
Trishala (mother of Mahavira) 71
Tvergastein (mountain hut) 8

Umasvati, *Tattvartha Sutra* of 38, 55–6, 65
Utsarpini (progressive half-cycle of time and the universe) 27

Vardhaman Gems (Jaipur, India) 7–8, 69–71, 72–3
Veerayatan movement 6–7, 33, 50, 60, 62–4, 68: development strategy 63, 66; pragmatic and realist objectives of 65–7; and self-sufficiency 66

vegetarianism/veganism 1, 25, 75
vidya (understanding or enlightenment) 13, 17
Vratas (five vows) 24–6, 27, 37, 39, 50–1, 54, 56–7, 61, 67–8

'web of life' (concept of) 2, 3, 5, 6, 13, 23
western thought: appeal of Anekantavada 38–40, 45, 51; Asian philosophies' influence in West 46, 47–8; classical liberalism 22; compartmentalising of knowledge 28–9, 54; conventional 'left'/'right' division 34, 46, 47; conventional western logic 5, 42, 54; Jain challenge to 29; Jain independence from 4–5; Jain openness to 2, 29; and Jain radical individualism 3, 22; orthodox economic models 66–7; *see also* environmental politics; green thought, western

Yin and Yang (in Chinese thought) 41
yoga 13, 14
Young Jains (organisation in UK) 75

Zen Buddhism 47

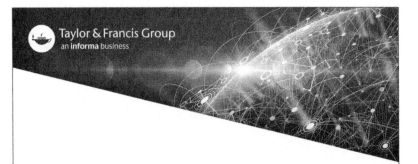

Printed in Great Britain
by Amazon